Transport and Receptor Proteins of Plant Membranes

Molecular Structure and Function

Transport and Receptor Proteins of Plant Membranes

Molecular Structure and Function

Edited by

D. T. Cooke

and

D. T. Clarkson

University of Bristol
Department of Agricultural Sciences
AFRC Institute of Arable Crops Research
Long Ashton Research Station
Bristol, United Kingdom

Springer Science+Business Media, LLC

Library of Congress Cataloging-in-Publication Data

Transport and receptor proteins of plant membranes : molecular
   structure and function / edited by D.T. Cooke and D.T. Clarkson.
        p.   cm.
    "Proceedings of the Twelfth Long Ashton International Symposium on
Transport and Receptor Proteins of Plant Membranes: Molecular
Structure and Function, held September 17-20, 1991, in Bristol,
United Kingdom"--T.p. verso.
    Includes bibliographical references and index.
    ISBN 978-0-306-44221-6   ISBN 978-1-4615-3442-6 (eBook)
    DOI 10.1007/978-1-4615-3442-6
    1. Plant proteins--Congresses. 2. Membrane proteins--Congresses.
3. Plant translocation--Congresses. 4. Plant molecular biology-
-Congresses. I. Cooke, D. T. (David T.) II. Clarkson, David T.
QK898.P8T73 1992
581.87'5--dc20   92-41404
                                                                CIP

Proceedings of the Twelfth Long Ashton International Symposium on Transport and
Receptor Proteins of Plant Membranes: Molecular Structure and Function,
held September 17-20, 1991, in Bristol, United Kingdom

ISBN 978-0-306-44221-6

© 1992 Springer Science+Business Media New York
Originally published by Plenum Press, New York in 1992

PREFACE

This book presents the contributions of the main speakers to the 12th Long Ashton International Symposium, held at the University of Bristol from 17th to 20th September 1991. Many of the 160 delegates who attended presented posters, the abstracts of which have been published separately and reflected the vigour, excitement and originality of the presentations.

The identity and molecular structure of solute of membrane proteins are of interest in themselves but may be seen as only a step towards broader questions about the regulation of their activity and their integration into the life of cells and organisms. This symposium was held at a time when exciting progress was being made on the molecular biology of some transporters, ion channels and hormone receptors. The contributions reflect this progress in different degrees. Some problems are less tractable than others and the mixture of success stories and hopeful aspirations was a deliberate choice by the organisers. At the symposium those who were still confronted with great difficulties were listened to sympathetically and with constructive interest. It is hoped that their papers will be read in the same way.

During the symposium we discussed the nature of many transport proteins and receptors, as well as the existence of additional regulatory proteins such as protein kinases, G-proteins and calmodulin. We could say that up to twenty proteins were considered in any detail. While this may sound like a considerable achievement, the scale of our ignorance can be illustrated by considering that on a typical silver-stained, two-dimensional polyacrylamide gel of purified plant plasma membrane polypeptides, between 150 and 200 spots are routinely shown; radiolabelled membranes frequently give even more, indicating how much membrane protein there is, the function of which we have no conceptual framework. Therefore, it would be rash to imagine that we know more than a little about how solute transport is brought about and regulated.

The papers in this volume start on reasonably firm ground, with proteins which can be isolated and, in some cases be re-constituted, after purification. Furthermore, the genes of some of these proteins have been cloned and their amino-acid sequences deduced. The role of these proteins in cellular physiology is clear and is of general importance. The ground becomes less certain as we move through the secondary metabolite and ion carriers, to channels and receptor proteins. This is, of course, largely determined by quantitative factors. Transporters (permeases) and channels facilitate transport at much greater rates than the primary transport enzymes, and are consequently much less abundant. Some of the same factors probably account for the low abundance of receptor proteins, but in this case, the problems of identification are compounded by the fact that they respond to relatively weak signals. Thus, using classical biochemical techniques, the ability to provide the correct signal coupled with the transient nature of receptors makes them very difficult to locate with any degree of certainty, and presents the researcher with yet another set of problems. However, with the introduction of molecular biological techniques rapid strides are being made in this field.

The final section considers the way in which proteins are targeted at, and inserted into their appropriate membranes. Recently, exciting advances have been made in this relatively new field and the contributions to the symposium reflected this.

Throughout this volume we have added our commentary to introduce subject areas and to reflect the main features of the discussion that took place during the symposium sessions, and in some cases the content of posters. The objective has been to link, where possible, the contributions and to air the concerns which they aroused in the audience.

Many thanks are due to all of the contributors for their excellent co-operation in bringing these proceedings so quickly to the press. The editors would like to take this opportunity to thank all those who helped with the organisation of the symposium; Harry Anderson, David Coupland, Sharon Child, Christine Cooke, Janette Knights, Ann Belcher, Bob Harvey, Ken Williams, Nick Smith, and Linda Hughes for typing the manuscript. The editors would also like to thank Richard Hooley and Jonathan Napier for their help in preparing the introductions to the final sections.

David T. Cooke
David T. Clarkson

CONTENTS

Primary Active Transport Proteins

Introduction

The P-type ATPases play crucial roles in keeping the cytoplasmic activities of H^+ and Ca^{2+} very much below their thermodynamic equilibria with the surroundings. The plasma membrane is permeable to both ions and without such ion pumps the cytoplasm, with its elaborate biochemical machinery, would become rapidly overwhelmed by phytotoxic levels of Ca^{2+} and pH values outside the permitted range of many enzymes. In this chapter authors concentrate on two kinds of P-type ATPase, the proton and calcium pumps, with an additional report on the novel H^+-translocating pyrophosphatase of the tonoplast membrane which may also play a role in cytoplasmic pH regulation. The locations of these ATPases and the PPase are illustrated in Figure 1.

In contrast to the ATPases of organelles (F_oF_1-types and vacuolar-type) the P-type ATPases are simple in structure, consisting of a single polypeptide chain with a molecular weight of 100-140 kDa. The functional enzyme may be a dimer. The success of research on these proteins has depended on their abundance, their retention of enzymic activity during membrane isolation and the fact that certain domains are highly conserved. Thus, the ATP-binding domain is very highly conserved in P-type ATPases of diverse origin, some of which were among the first membrane proteins to be isolated and characterised, e.g. the mammalian erythrocyte Na/K^+ pump. Because oligonucleotide probes to this conserved region are readily prepared, the task of searching for mRNA's and genes in hitherto unexplored plant species has been greatly simplified and progress in recent years has been quite impressive.

Deduced amino acid sequences, sometimes directly confirmed by amino acid analysis, give a picture of P-type ATPase structure which is becoming quite familiar. Much of the sequence is buried in the membrane, there being from 6-10 membrane-spanning α-helices. Very little of the sequence is on the external face of the membrane, but on the cytoplasmic side there are long loops of hydrophilic residues. These comprise the domains where ATP is bound and the regulatory phosphorylation sites. Some, like the ATP-binding region are very highly conserved while the non-conserved regions must contribute to the specific properties of ATPases of different origin.

The broad outline of the reaction mechanism of P-type ATPases was understood well in advance of these structural features. It is now clear, however, that transport of "substrate" ions across the membrane depends on properties of two kinds:

(i)   The assembly of the membrane-spanning helices to form a pore through which transport can occur.

(ii)   Changes in the folding of the polypeptide chain leading to ion-binding sites being alternately exposed to the milieux on either side of the membrane.

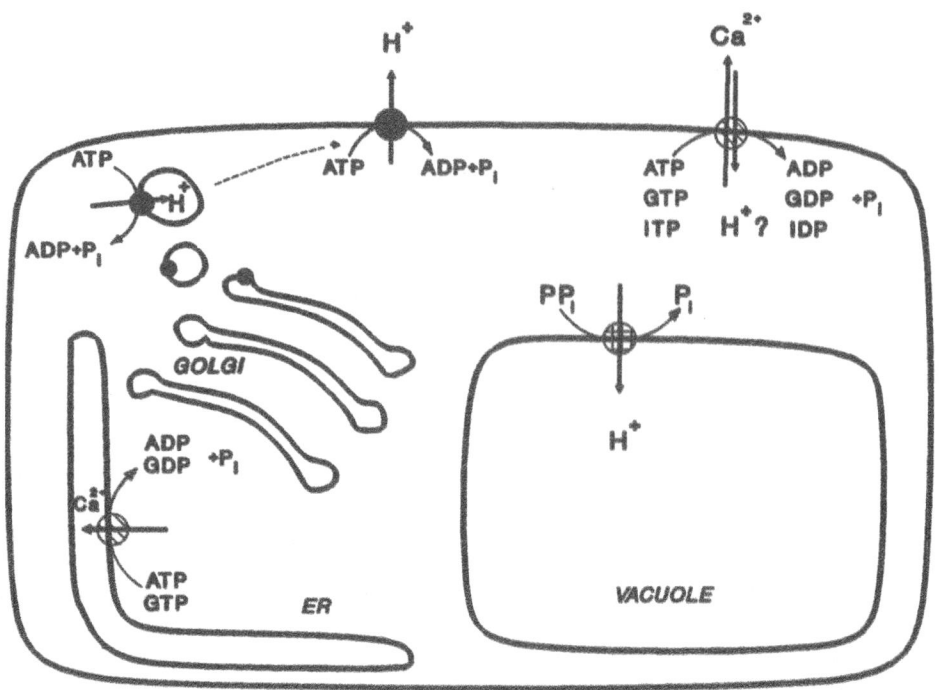

Figure 1 The location of the ATPases and PPase described in the contributions which follow (Produced by A M Ramon).

With some modification these generalisations could be extended to cover secondary ion and metabolite transporters and channels.

To repeat this is to give the impression that we know more than we actually do. As far as the plant P-type ATPases are concerned the above is a 'concept' rather than a description. Clearly, in polypeptides of great length, certain residues are more important than others in determining catalytic and transport functions. The biochemist may seek to probe these crucial residues by direct chemical modification (Briskin) while the molecular biologist may delete or change them by site-directed mutagenesis. In higher plants the latter process has scarcely begun but in yeasts much has been learnt already, particularly about the nature of the regulatory domain. Now that heterologous expression of higher plant genes in yeast can be routinely achieved (Sauer) a possibility exists for studying the function of mutated genes for the H^+-pump. Recently a major difficulty in this work has been overcome by expressing mutated genes in *sec* mutants of yeast, where vesicles containing the defective pump accumulate in the cytoplasm but do not fuse with the plasma membrane. Disruption of the cells yields ready-made vesicles in an appropriate orientation so that catalytic and transport functions can be assayed (Nakamoto, R.K., *et al.*, 1991, J. Biol. Chem., **266**, 7940-7949). Previously many mutated genes have proved to be lethal on their own, so that they could be expressed only in cells which also expressed the wild-type gene, the experimenter being left to apportion effects to the two systems.

A complete understanding of the molecular mechanisms of membrane-associated processes will depend on detailed three-dimensional structural information. Such information can come from x-ray crystallography, but low abundance, amphipathic proteins are unpromising material from which to obtain crystals. Alternative approaches are necessary, therefore, and may come from high resolution NMR studies, particularly of synthetic α-helices from the membrane spanning regions which are known to associate with one another in non-aqueous solvents.

It is clear that studies of the H^+-ATPase entered a new phase with the discovery that there are several genes (Boutry, M. *et al.*, 1989, Biochem. Biophys. Res. Commun., **162**, 567-574; Ewing, N.N., *et al.*, 1990, Plant Physiol., **94**, 1874-1881) within a given organism which code for similar versions of the proton pump (Sussman). What are we to make of these subtle variations? Sometimes this complexity is related to tissue specific expression which might imply differences in the regulation of pump activity in different cell types. Perhaps small variations in structure may have significance in the interactions of the H^+-ATPase with other proteins in the membrane. This suggestion opens a new vista of non-chemiosmotic energy transfers, where a change in the conformation of one protein may directly affect that of neighbouring one. Thus different versions of the pump may interact with different proteins. In discussion it was pointed out that the lipid composition of the plasma membrane may vary between cell types and that lipids can regulate ATPase activity (Palmgren, M.G., *et al.*, 1990, J. Biol. Chem., **265**, 13423-13426; Kasamo, K. and Nouchi, I., 1987, Plant Physiol., **83**, 823-828).

The molecular biology of the Ca^{2+}-ATPases from plants has not advanced very far, but there is already a suggestion that several genes encode this activity (Sussman, this volume). Given the special role of cytoplasmic calcium in cell signalling and the wide range of responses in different cell types, it would not be surprising if a 'plethora' of calcium pumps came to light in the near future.

A Plethora of Plant Plasmalemma Proton Pumps

Michael R. Sussman

Department of Horticulture
University of Wisconsin
1575 Linden Drive
Madison, WI 5306, USA

Strongly hydrophobic integral membrane proteins are difficult to purify and characterise using conventional biochemical techniques. Eight years ago my co-workers and I initiated a research programme using molecular biological techniques to help elucidate the structure and function of important proteins in the plasma membrane of plant cells. We first turned out attention to the H^+-ATPase, since this enzyme is easy to assay and is one of the more abundant transport proteins in this membrane (Surrowy and Sussman, 1987).

Our initial aim was to isolate large amounts of the enzyme needed to obtain protein sequence information. We decided to use roots derived from 6-day-old etiolated oat seedlings because this tissue could be grown and harvested in large amounts and, when compared to many other easily obtainable plant tissues, gave high values for extractable ATPase activity.

Studies in other laboratories had already established that the major plasma membrane ATPase in these cells is a H^+-ATPase which is inhibited by vanadate and diethylstilbesterol, but is resistant to compounds which inhibit ATPase from other subcellular fractions: 5 mM NaN_3 for the mitochondrial H^+-ATPase, 100 mM KNO_3 for the vacuolar H^+-ATPase and 0.1 mM molybdate for a non-specific phosphatase that contaminates the plasma membrane vesicle fraction (Vara and Serrano, 1982). Although potassium usually elicits a stimulation of ATPase activity *in vitro*, this stimulation is probably non-specific and does not reflect the movement of potassium by this enzyme. The purified enzyme contains a catalytic subunit of ca. $M_r = 100,000$ which forms a phosphorylated intermediate (a base labile phospho-aspartyl bond) during the reaction cycle. The molecular weight and catalytic properties of this enzyme align it with a group of 'P-type' ATPases readily distinguishable from the F_oF_1-type ATPase of bacteria, mitochondria and chloroplast membranes, and the 'V-type' ATPase of the tonoplast.

Transport and Receptor Proteins of Plant Membranes
Edited by D.T. Cooke and D.T. Clarkson, Plenum Press, New York, 1992

5

The protocol developed in my laboratory allowed one individual to obtain, in less than 8 hours, ca. 20-30 mg of purified plasma membrane vesicles, starting with 0.5 kg of oat roots. The purified membranes displayed an H$^+$-ATPase specific activity of 1-2 μmoles/minute/mg protein. Similar specific activities were obtained when plasma membrane vesicles were purified by either a discontinuous sucrose density gradient centrifugation or a two-phase aqueous dextran partition procedure. A careful balance sheet of H$^+$-ATPase activity obtained at each step of the purification demonstrated that the "purified" plasma membrane vesicles were enriched ca. 30-fold during purification and the overall yield was fairly good, i.e. over 50% (Katz, 1989). This indicated that, based on Lowry protein measurements, ca. 3% of the total cell protein was present in the plasma membrane of oat roots.

By treating the plasma membrane vesicles with low concentrations of a detergent such as deoxycholate or Triton X-100, ca. 50% of the membrane protein becomes solubilised and no longer pellets with the vesicles. In contrast, no ATPase is solubilised with this treatment and the specific activity of ATPase in the pelleted detergent-washed vesicles is approximately doubled. Most important, this simple wash procedure removes many loosely bound contaminating proteins which, on SDS-PAGE, display a denatured molecular weight similar to that of the ATPase (100 kilodaltons). By accumulating 200 mg of plasma membrane vesicles from many days of oat root extractions and then washing with detergent, we obtained a preparation which on SDS-PAGE (8% (w/v) acrylamide) showed a distinct polypeptide at $M_r = 100,000$ which was one of the darkest high molecular weight Coomassie-stained bands on the gel (cf. Surowy and Sussman, 1986). We were able to mimic this electrophoretic SDS-PAGE separation via chromatographic molecular sieving with a BioGel A1.5M column chromatography in the presence of 1% (w/v) lithium dodecyl sulfate (lithium was used in order to run the column in the cold room - sodium salts of SDS are more susceptible to precipitation in the cold). After solubilising the 100 mg of detergent-washed vesicles in a small volume of 10% (w/v) lithium dodecyl sulfate, the sample was applied to this column. SDS-PAGE analysis of eluted fractions showed several fractions in which the ATPase was almost completely pure. Lowry protein measurements revealed that ca. 1 nanomole of ATPase was obtained from the BioGel A1.5M column, indicating that the original 200 mg of plasma membrane vesicles contained ca. 0.5% (w/v) (i.e. 1 mg) of ATPase protein. In other words, in the crude root homogenate, the ATPase polypeptide represented ca. 0.5% of 3%, or 0.0015% of the total cell protein. Overall, this indicates that the ATPase, which is one of the more abundant proteins in the plant plasma membrane, represents only one out of 10,000 proteins, which is similar to the abundance observed for ATPase cDNA clones in a total cell cDNA library (see below).

After treating the purified ATPase polypeptide with iodoacetamide to alkylate cysteine residues, SDS was removed from the sample by adsorption to "Bio-Beads". The protein was then digested overnight with trypsin and applied to a reverse phase HPLC column. Tryptic peptides were resolved with an acetonitrile gradient and eight random peptides were chosen for automated Edman degradations on a gas-phase protein sequencer (Schaller and Sussman, 1988).

Protein sequence data provided from the above experiment, plus that known from two conserved short sequences in the active site of all P-type ATPases allowed us to design and synthesise an oligonucleotide specific for the plant plasma membrane H$^+$-ATPase. We screened an oat cDNA library with this oligonucleotide and obtained one representing a ca. 2/3 full-length cDNA clone, at a frequency of ca. 1/10,000. This oat H$^+$-ATPase clone was then used to screen *Arabidopsis thaliana* cDNA libraries and two different cDNA clones were obtained. When one full-length clone was completely sequenced it was striking to observe at least 90% identity between the amino acid sequence of the random oat peptides and the DNA-derived amino acid sequence of the *Arabidopsis* clone (Harper, Surowy and Sussman, 1989).

At this point we had two important pieces of information. First, we were able to pinpoint in the cDNA sequence, the sequence of all eight ATPase tryptic peptides, confirming that the ATPase polypeptide was indeed homogeneous and purified free of contaminating non-ATPase polypeptides. A second conclusion was that the ATPase was highly conserved between monocots and dicots, a conclusion supported by earlier immunological analyses (Surowy and Sussman, 1986).

From 1988 to 1990, my laboratory determined the full-length sequence of cDNA for two *Arabidopsis* ATPase isoforms (Harper, Manney, DeWitt, Yoo and Sussman, 1990), and Serrano's laboratory determined the sequence for a third (Pardo and Serrano, 1989). We called each of these isoforms AHA-1, AHA-2 and AHA-3 (after *Arabidopsis* H^+-ATPase). A fundamental question which immediately came to mind was, are there only three isoforms? Low stringency hybridisations gave multiple bands on Southern Blots, but it is difficult to be sure that the hybridising bands are true ATPase genes, not just spurious matchings. As an alternative approach, we synthesised two oligonucleotides which could act as primers for a polymerase chain reaction (PCR) to amplify all known eukaryotic P-type ATPase genes. These two primers were directed against two active site sequences which, in AHA1-3, were separated by ca. 800 bp ($\sim$ 300 amino acids). The oligonucleotides were highly degenerate (>1000 fold) but a PCR is much more forgiving that library screening when it comes to potentially negative effects of degeneracy. With *Arabidopsis* genomic DNA as a template, we observed multiple bands of 800 bp and greater, when the PCR reaction was analysed on ethidium bromide-stained agarose gels. DNA extracted from all of these bands were cloned into a suitable plasmid and the region between the primers was sequenced. This analysis revealed many diagnostic P-type ATPase sequences *between* the two highly conserved primer regions, indicating that the clones probably arose from true ATPase gene templates, rather than PCR artefacts. In order to test this further, we were fortunate to find an intron whose position was highly conserved in the amplified region. As expected, the intron sequences for each clone showed no homology with each other. We then prepared intron-specific radioactive probes and verified that each of the PCR-derived P-type ATPase clones gave dark, unique bands on Southern Blots performed with *Arabidopsis* genomic DNA. We have since isolated these various AHA genomic clones by screening *Arabidopsis* cosmid libraries with these gene-specific probes. The final conclusion of this analysis is that the first three ATPase cDNA's which we sequenced are members of a larger gene family, containing at least ten different gene isoforms (AHA1-10). All of these 10 isoforms encode a P-type ATPase with 80-90% amino acid identity to the original AHA1-3 isoforms. There are, in addition, two other P-type ATPase isoforms which only display 60-80% amino acid identity to the first three AHA clones. We have termed these last two genes, AXA1 and AXA2, since the divergence of sequence suggests that ATPases with different catalytic functions (e.g. a Ca^{2+}-translocating ATPase) may be encoded. We are in the process of (1) mapping the chromosomal location and sequencing each of these *Arabidopsis* P-type ATPase genes and, (2) determining when and where in the plant each is expressed.

In mammals, there is likewise a large gene family encoding the plasma membrane P-type Na^+,K^+-ATPase (see Sussman and Harper, 1989). The different isoforms seem to be expressed in different specialised transport tissues. For example, two genes are reserved for expression in brain cells, a third is seen mainly in kidney and another appears to be developmentally and/or hormonally regulated. It is thus reasonable to ask whether the multiplicity of *Arabidopsis* plasma membrane P-type ATPases serve a similar function, i.e. for expression of unique isoforms in specialised 'transport' cells of higher plants. As described below, the preliminary indication from our studies is a tentative "yes".

The great similarity of the ten AHA cDNA's makes it difficult to develop gene-specific RNA *in situ* hybridisation probes. As an alternative, we have utilised the reporter gene encoding glucuronidase (GUS) in a translational fusion to ATPase genomic clones. The GUS gene is fused in-frame within a few amino acids at the 3' end of the ATG start codon of the ATPase. By incorporating 3-5 kilobases of DNA sequence 5' upstream from this region, we hope to determine the spatial and temporal pattern of expression directed by this promoter in the normal plant. Transgenic tobacco and *Arabidopsis* plants are generated with these fusions, and GUS activity is determined in histological thin sections throughout the plant. We have completed this analysis for AHA2 and AHA3, i.e. only two of the ten AHA isoforms and each shows a different pattern of expression. The AHA3 promoter directs GUS, in vegetative tissue, exclusively to phloem cells in vascular tissue throughout the plant (DeWitt, Harper and Sussman, 1991). This explains why, in Northern Blots, AHA3 mRNA is observed in both roots and shoots (Harper *et al.*, 1990). When high-resolution histological thin sections are made, GUS activity is observed only in sieve-tube cells and companion cells. GUS activity is found also in pollen grains and ovule tissue. Most striking was the observation of GUS activity in the funiculus, a structure analogous to the mammalian umbilical cord that nourishes the developing embryo. In summary, the AHA3 promoter seems to direct GUS expression to cells involved in sucrose transport.

Physiological experiments have suggested that sucrose transport in plant cells is mediated by a H^+-co-transport system. The activity of this symporter is especially important in phloem cells since this is the main conduit for sugar transport between sites of synthesis (i.e. sources such as leaves) and sites of utilisation (i.e. sinks such as roots or embryos). Although the ten P-type ATPases show a great deal ($\geq 80\%$) of overall amino acid identity, there are shorter regions which are unique for each protein. It is possible that these provide important catalytic differences which render a particular ATPase more suitable for catalytic functions in a given cell type. Although current dogma states that the H^+-ATPase is coupled only indirectly to the function of a sucrose symporter via the bulk protonmotive force, it is possible that the pump and carrier polypeptides are colliding within the plane of the membrane. There are recent reports of observations which do not fit purely chemiosmotic models for indirect coupling between pumps and carriers (Janoshazi and Solomon, 1989; Janoshazi, Sheifter and Solomon, 1989). If conformational couplings actually occur during *in vivo* transport functions, this suggests that specific pump sequences may be required for maximal interactions with specific carriers.

Reporter gene analysis with AHA2 is not yet published, but preliminary results reveal a pattern of expression unlike that seen with AHA3. Thus, with the AHA2 promoter fused to GUS in transgenic plants, reporter gene activity is observed only in roots and only in non-vascular tissue of this organ. The epidermal and cortical cells showed high GUS activity but xylem and phloem showed none. Northern Blot analysis confirms that AHA2 and AHA1 are predominantly expressed in roots. It is interesting to note that with plants grown under different conditions, AHA1 expression deviates from this pattern, in that high levels of expression are observed in shoots. This indicates that AHA1 expression may be regulated by hormones or other physiological factors. It is useful to note here that in *E. coli*, the plasma membrane K^+-ATPase expression is indicated by the loss of turgor that ensues during potassium starvation (Laimins, Rhoads and Epstein, 1981). It is well-known that mineral starvation causes large changes in transport functions of higher plants and a change in one ATPase isoform expression is a possibility worth examining. Finally, it is conceivable that expression of one of the ATPase isoforms is specifically activated by fusicoccin or auxin since an increase in the plasma membrane protonmotive force is an early effect of these growth stimulators. Further studies with clones DNA sequences derived from each of the AHA and AXA gene isoforms may provide important clues for the *in situ* function of the plasma membrane H^+-ATPase in higher plants.

A separate question under study in my laboratory for the past few years concerns the molecular mechanism by which the plant plasma membrane proton pump is regulated post-translationally. Electrophysiological and biochemical evidence over the past two years indicates that the ATPase catalytic activity is altered within 15-30 seconds after certain treatments. These include: (1) blue light on guard cells, (2) fusicoccin addition to almost any plant cell, (3) glucose addition to yeast cells and (4) 0.3 M mannitol addition to *Neurospora crassa*. All of these effects are only observed *in situ*, i.e. the ATPases of isolated membrane vesicles are unaffected by *in vitro* treatments. Several years ago, we found that the oat root ATPase is a phosphoprotein (P-ser and P-thr) *in vivo* and we hypothesised that kinase-mediated phosphorylation of the ATPase is the means by which the catalytic activity is altered post-translationally (Schaller and Sussman, 1987, 1988).

In order to test this hypothesis we decided first to characterise the kinase(s) responsible for the ATPase phosphorylation. We observed that when purified oat root plasma membrane vesicles are incubated with $\gamma[^{32}P]$-ATP, radioactivity is incorporated into the ATPase, at serine and threonine residues. The majority of this kinase-mediated ATPase phosphorylation was found to be strictly dependent on the presence of sub-micromolar concentrations of calcium.

We were very fortunate in this project to initiate a collaboration with Dr. Alice Harmon (University of Florida, Gainesville, FL). Dr. Harmon and co-worker were able to purify to homogeneity a calcium-dependent protein kinase (CDPK) from the cytoplasmic fraction. They prepared four monoclonal antibodies against the kinase and were able to obtain amino acid sequence data for several endoproteolytically-derived kinase peptides. By performing Western Blots with the kinase antibodies, we found that the oat root plasma membrane calcium-dependent protein kinase was 100% cross reactive with each of the four monoclonals. We observed further that the oat root plasma membrane CDPK showed a $M_r=61,000$ Western blot band which could form smaller sized bands (e.g. $M_r=55,000$, the size of the cytoplasmic soybean CDPK) during storage. Furthermore, we observed that when the oat root plasma membrane CDPK is treated with low concentrations of trypsin, it is proteolytically "clipped" to a smaller M_r (i.e. 61,000 to 58,000) and coincidently, it becomes less hydrophobic, as assayed by partitioning into Triton X-114. Finally, when oat roots are homogenised under conditions which minimise proteolysis, all anti-CDPK Western reactive material is found associated with microsomes and none is seen in cytoplasmic fractions (Schaller, Harmon and Sussman, in press).

In order to learn more about this kinase, we proceeded to utilise Dr. Harmon's peptide sequence data to clone the soybean and *Arabidopsis* CDPK gene, using techniques similar to those used in my laboratory to clone ATPase gene isoforms (Harper, Sussman, Schaller, Putnam-Evans, Charbonneau and Harmon, 1991). When the completed amino acid sequence of CDPK clones was analysed we observed a surprising result: the soybean CDPK, of $M_r \sim 55,000$, contained an amino terminal $M_r=35,000$ kinase catalytic domain, attached to a carboxy terminal $M_r=20,000$ 'regulatory' domain with four calcium binding sites (EF hands). A GENBANK database search revealed that the CDPK catalytic domain most closely resembles animal *calmodulin-dependent protein kinases* and the CDPK regulatory domain most closely resembles *calmodulin*. We have been able to express the

CDPK clone in *E.coli*, producing CDPK activity in *E.coli* extracts using an expression plasmid. Most interestingly, the expressed clone displays lipid-stimulated activity using histone as substrate. In addition, the oat root plasma membrane H^+-ATPase is an excellent substrate for the expressed CDPK clone. We have isolated two CDPK gene isoforms in *Arabidopsis thaliana* and are in the process of characterising their molecular structure and patterns of expression.

The lipid activation of the expressed cloned gene is particularly interesting in light of a similar activation of a CDPK-like enzyme partially purified from oat root plasma membranes. No activity is found with pure phosphatidylserine and phosphatidylcholine, but in the presence of a crude soybean lipid preparation the enzyme is highly active. It is possible that previous reports in the literature of a protein kinase C-like enzyme in crude extracts of higher plants was actually measuring CDPK activity, rather than a true plant protein kinase C. As yet, no DNA clone, encoding plant protein kinases, which resembles protein kinase C has been found. Therefore, it is possible that a CDPK-like enzyme with lipid-stimulated properties may act as a central component of plasma membrane transduction events unique to plants. Further biochemical studies with the plant enzyme are needed to clarify whether this is idle speculation or fact.

In vitro experiments demonstrate that the M_r-100,000 H^+-ATPase is not the only substrate for plasma membrane CDPK. CDPK is the predominant protein kinase in crude extracts of all plant cells we have examined. The identification of its *in situ* plasma membrane substrates and their role in plant signal transduction remains an important and fascinating subject for future investigations.

REFERENCES

DE WITT, N.D., HARPER, J.F., and SUSSMAN, M.R., 1991. Evidence for plasma membrane proton pump in phloem cells of higher plants. *The Plant Journal*, 1, 121-128.

HARPER, J.F., MANNEY, L., DE WITT, N.D., YOO, M.H., and SUSSMAN, M.R., 1990. The *Arabidopsis thaliana* plasma membrane H^+-ATPase multigene family: genomic sequence and expression of a third isoform. *Journal of Biological Chemistry*, 265, 13601-13608.

HARPER, J.F., SUROWY, T.K., and SUSSMAN, M.R., 1989. Molecular cloning and sequence of cDNA encoding the plasma membrane proton pump (H^+-ATPase) of *Arabidopsis thaliana*. *Proceedings of the National Academy of Sciences, USA*, 86, 1234-1238.

HARPER, J.F., SCHALLER, G.E., SUSSMAN, M.R., PUTNAM-EVANS, C., CHARBONNEAU, H., and HARMON, A.C., 1991. A calcium-dependent protein kinase with a regulatory domain that is similar to calmodulin. *Science*, 252; 951-954.

JANOSHAZI, A., SEIFTER, J.L., and SOLOMON, A.K., 1989. Interactions between anion exchange and other membrane proteins in rabbit kidney medullary collecting duct cells. *Journal of Membrane Biology*, 112, 39-49.

JANOSHAZI, A., and SOLOMON, A.K., 1989. Interaction among anion, cation and glucose transport proteins in the human red cell. *Journal of Membrane Biology*, 112, 25-37.

KATZ, D., 1989; Studies on the plasma membrane H^+-ATPase of oat roots - preparation and assay, cytological localisation and sulfhydryl chemistry. Ph.D. thesis, University of Wisconsin, Madison, WI.

LAIMINS, L.A., RHONDS, D.B., and EPSTEIN, W., 1981. Osmotic control of kdp operon expression in *Escherichia coli*. *Proceedings of the National Academy of Sciences, USA*, 78, 464-468.

PARDO, J.M. and SERRANO, R., 1989. Structure of plasma membrane H^+-ATPase gene from the plant *Arabidopsis thaliana*. *Journal of Biological Chemistry*, 264, 8557-8562.

SCHALLER, G.E., HARMON, A.C., and SUSSMAN, M., 1992. Characterization of calcium- and lipid-dependent protein kinase associated with the plasma membrane of oat. *Biochemistry*, in press.

SCHALLER, G.E., and SUSSMAN, M.R., 1987. Kinase-mediated phosphorylation of the oat plasma membrane H^+-ATPase. In *Plant Membranes: Structure, Function and Biogenesis. UCLA Symposium on Molecular and Cellular Biology, New Series, Volume 63.* Eds C. Leaver and H Sze.

SCHALLER, G.E. and SUSSMAN, M.R., 1988. Phosphorylation of the plasma membrane H^+-ATPase of oat roots by a calcium-stimulated protein kinase. *Planta*, **173**, 509-518.

SCHALLER, G.E. and SUSSMAN, M.R., 1988. Isolation and sequence of tryptic peptides from the oat plasma membrane H^+-ATPase. *Plant Physiology*, **86**, 512-516.

SUROWY, T.K. and SUSSMAN, M.R., 1987. Molecular cloning of the plant plasma membrane H^+-ATPase. *Plant and Soil*, **99**, 185-196.

SUROWY, T.K., and SUSSMAN, M.R., 1987. Immunological cross-reactivity and inhibitor sensitivities of the plasma membrane H^+-ATPase from plants and fungi. *Biochemica et Biophyica Acta*, **848**, 24-34.

SUSSMAN, M.R. and HARPER, J.F., 1989. Molecular biology of the plasma membrane of higher plants. *The Plant Cell*, **1**, 953-960.

SUSSMAN, M.R. and SOROWY, T.K., 1987. Physiology and molecular biology of membrane ATPases. In *Oxford Surveys of Plant Molecular and Cellular Biology*, Volume 4. pp 47-71. Ed B.J. Miflin. Oxford University Press, Oxford. (Review)

VARA, F., and SERRANO, R., 1982. Partial purification and properties of the proton translocating ATPase of plant plasma membranes. *Journal of Biological Chemistry*, **252**, 12826-12830.

Studies on the Reaction Mechanism and Transport Function of P-type ATPases Associated with the Plant Plasma Membrane

Donald P. Briskin, Swati Basu and InSun Ho

Department of Agronomy
University of Illinois
Urbana, Illinois, 61801, USA

1.   Introduction

Although it was once common to refer to the ATP hydrolytic activity associated with the plasma membrane as "the plasma membrane ATPase", it is now apparent that at least two distinct ATP hydrolysing enzymes are associated with this membrane and involved in primary active transport. From studies conducted over the past two decades using intact plant tissues, isolated cells, membrane vesicles and reconstituted enzyme preparations, a plasma membrane H^+-ATPase has been characterized which functions in coupling ATP hydrolysis to H^+ extrusion at the cell surface (Briskin, 1990a and references therein). Through this activity, the H^+-ATPase has an important physiological role in establishing an inwardly-directed proton electrochemical gradient across the plasma membrane which can provide the driving force for the transport of numerous solutes and metabolites via H^+/solute symports, antiports or membrane potential-driven uniports and channels (Sanders and Slayman, 1989). In more recent studies conducted over the past 5 years, a second ATP hydrolysing enzyme, the Ca^{2+}-ATPase, has also been shown to be associated with the plasma membrane and involved in mediating primary Ca^{2+} efflux from the plant cell in a manner independent of the ΔpH- or $\Delta\psi$-linked transport processes driven by the plasma membrane H^+-ATPase (Briskin 1990b). The activity of this Ca^{2+}-translocating ATPase may be important for maintenance of the low cytoplasmic Ca^{2+} concentration ($0.1~\mu M$) and a steep Ca^{2+} gradient at the plasma membrane (10^4-fold difference) required for the function of this divalent cation as a second messenger in the regulation of cellular processes (Marmé, 1989). Activation of the plasma membrane Ca^{2+}-ATPase by calmodulin would further indicate that its activity is modulated in accordance with cytoplasmic Ca^{2+} concentration levels (Briskin, 1990b; Marmé, 1989).

Whilst the plasma membrane H^+-ATPase and Ca^{2+}-ATPase are involved in different physiological functions, they are similar in general structure and mechanism; being P-type

Transport and Receptor Proteins of Plant Membranes
Edited by D.T. Cooke and D.T. Clarkson, Plenum Press, New York, 1992

13

transport ATPases (Pederson and Carafoli 1987; see next section). Such enzymes are both mechanistically and structurally different from the V-type ATPases associated with plant vacuoles, secretory vesicles and animal lysosomes (Stone, Crider, Südhof, and Xie, 1990) and from the F-type ATPases associated with mitochondria, chloroplasts and prokaryotes (Futai, Noumi, and Maeda, 1989). In this paper, aspects of the enzyme mechanism, structure and transport function of these two plasma membrane-associated P-type transport ATPases will be considered.

2.   Reaction Mechanisms with Phosphorylated Intermediates

A characteristic feature of P-type transport ATPases, such as the plasma membrane H^+-ATPase and Ca^{2+}-ATPase, is their formation of a covalent phosphorylated intermediate during the course of substrate hydrolysis (Pederson and Carafoli, 1987). The mechanisms for substrate hydrolysis can minimally be described according to a two-step process:

$$E + ATP \text{-----}> E\text{-}P + ADP \tag{1}$$

$$E\text{-}P + H_2O \text{-----}> E + P_i \tag{2}$$

which would appear to represent an Ordered Uni Bi mechanism in the simplest sense (Rudolph, 1983). It should be noted that an ordered product release of ADP before Pi has only been confirmed for the plasma membrane H^+-ATPase using transient state kinetic methods (Briskin, 1988). Based upon both its acid stability and discharge by hydroxylamine, the protein phosphate bond in the plasma membrane H^+-ATPase (Briskin, 1990a and references therein) and Ca^{2+}-ATPase (Williams, Schueler and Briskin, 1990; Briars and Evans, 1989) has been shown to be an acylphosphate typically observed for P-type transport ATPases (Pederson and Carafoli, 1987). By analogy to other transport ATPases, this bond would likely involve an aspartyl residue in the active site (Pederson and Carafoli, 1987) although this has only been shown experimentally for the plasma membrane H^+-ATPase (Briskin and Poole, 1983a; Walderhaug, Post, Saccomani, Leonard, and Briskin, 1985).

Formation of an acyl-linked phosphorylated intermediate by the plasma membrane H^+-ATPase and Ca^{2+}-ATPase would also explain the observed vanadate sensitivity of these two enzymes (Briskin, 1990a; 1990b; and references therein). The reaction involving hydrolysis of the phosphorylated intermediate (*rxn 2*) by H_2O would likely occur via an "in line" nucleophilic attack on the protein-bound phosphate group involving either a penta co-ordinate transition state or enzyme intermediate (Knowles, 1980). As pentavalent vanadium (orthovanadate) can easily adopt a stable trigonal bi-pyramidal structure resembling this transition state or enzyme intermediate (Macara, 1980; Chasteen, 1983), tight binding of this inhibitor would lock the enzyme in this conformation state preventing subsequent substrate binding and hydrolysis (Figure 1).

It has been a general observation, from the study of several P-type transport ATPases, that the ion to be translocated from the cytoplasmic face of the enzyme to an extra-cytoplasmic domain (cell exterior or lumen of an organelle) is a required ligand in reactions associated with phosphoenzyme formation (*overall rxn 1*). Hence, what is observed in kinetic studies is an ion-dependent phosphorylation of the ATPase. This has been best shown for the interaction of Na^+ in the mechanism of the animal cell Na^+,K^+-ATPase (Karlish, 1989) and for Ca^{2+} in the mechanism of both the animal plasma membrane and endoplasmic reticulum Ca^{2+}-ATPases (Schatzmann, 1989). The enhancement of phosphoenzyme formation for the plant plasma membrane H^+ATPase at low pH (Briskin, 1988; Briskin, 1986) and a Ca^{2+}- dependence for phosphorylation of the plant plasma membrane Ca^{2+}-ATPase (Briars and Evans, 1989) would again appear consistent with this scheme.

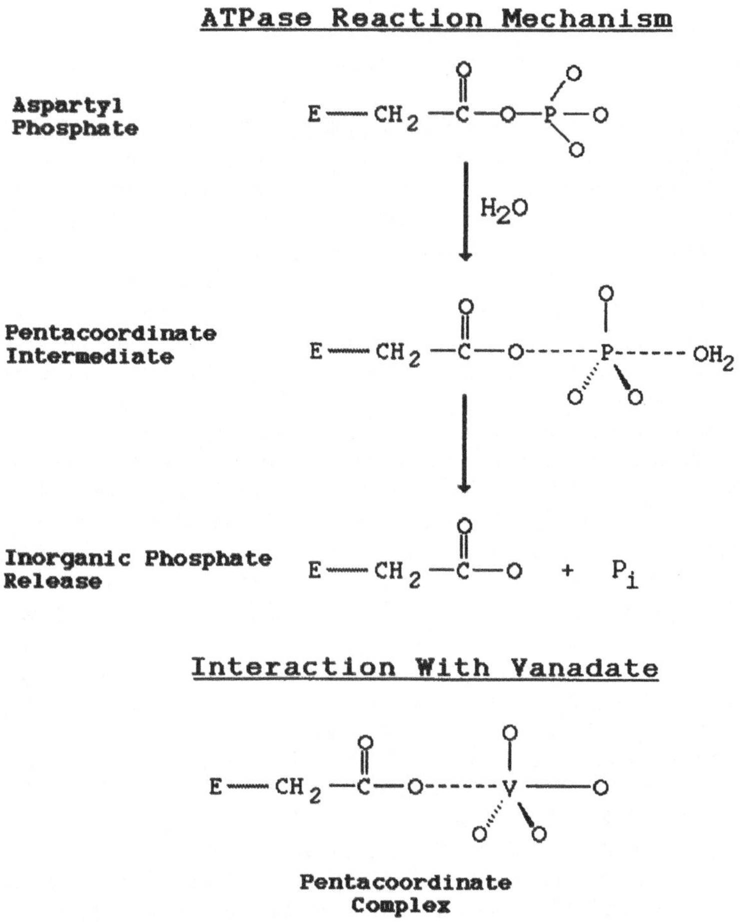

ATPase Reaction Mechanism

Figure 1 Formation of a penta co-ordinate intermediate during hydrolysis of the acyl phosphate and its relationship to vanadate inhibition.

The plasma membrane H^+-ATPase (Briskin, 1990a and references therein) and Ca^{2+}-ATPase (Kasai and Muto, 1990; Malatialy, Greppin, and Penel, 1988) require the presence of Mg^{2+} for activity and such a requirement may be due to the Mg:ATP complex representing the true substrate for phosphoenzyme formation (Briskin, 1990a and references therein). Although it has been shown that dephosphorylation of the plant plasma membrane H^+-ATPase does not require Mg^{2+} (Briskin and Poole, 1983b, Briskin, 1986), it is uncertain if this is also true for the plasma membrane Ca^{2+}-ATPase. For both the animal plasma membrane Ca^{2+}-ATPase (Schatzmann, 1989; Carafoli, 1991) and sarcoplasmic reticulum Ca^{2+}-ATPase, Mg^{2+} has the additional role of accelerating enzyme dephosphorylation.

3.   Studies with Protein Modification Reagents

To understand ultimately how the plasma membrane H^+-ATPase and Ca^{2+}-ATPase can couple ATP hydrolysis to primary ion transport, it will be necessary to identify active site amino acid residues involved in the catalytic/transport mechanism and then elucidate their specific roles in this process. A powerful technique for initial identification of these "essential" amino acid residues involves the use of protein modification reagents (Eyzraguirre, 1987 and references therein). Under appropriate reaction conditions, such reagents react with specific amino acid types to form a covalent bond. If it is assumed that this does not result in a major change in protein conformation, then any loss of enzyme activity can be attributed to involvement of that particular amino acid in the mechanism of the protein (Eyzraguirre, 1987). Protection against this activity loss by inclusion of the substrate during treatment of the enzyme with the modification reagent would also be consistent with an active site location for the derivatised amino acid. When combined with the information from molecular studies currently being conducted on the plasma membrane H^+-ATPase and Ca^{2+}-ATPase, work with protein modification reagents may prove useful for providing clues as to the possible location of essential amino acids and catalytic domains within deduced linear amino acid sequences for these enzymes.

For the plant plasma membrane H^+-ATPase, work with protein modification reagents has allowed identification of several essential amino acids. For a few of these, functional roles can be suggested from analogy to other well characterized P-type transport ATPases and soluble enzymes. Involvement of an essential lysine residue in the mechanism of the plasma membrane H^+-ATPase has been indicated from the observation that this enzyme is inhibited by fluorescein isothiocyanate (FITC) (Gildensoph, 1989). Inhibition by this reagent has also been observed for the animal cell Na^+,K^+-ATPase (Farley, Tran, Carilli, Hawke, and Shively, 1984), sarcoplasmic reticulum Ca^{2+}-ATPase (Pick and Karlish, 1980) and fungal plasma membrane H^+-ATPase (Pardo and Slayman, 1988). Biochemical and molecular studies with these enzymes has suggested a role for the FITC-derivatisable lysine in interacting with the adenine "handle" portion of the nucleoside triphosphase substrate (Cantley, Carilli, Farley, and Perlman, 1982). A further similarity to these other P-type ATPases (Kasher, Allen, Kasamo, and Slayman, 1986; Depont, Schoot, VanProoigen-VanEeden, and Bonting, 1977; Murphy, 1976) is the observed inhibition of the plant plasma membrane H^+-ATPase by arginine-derivatising reagents such as 2,3-butanedione (Gildensoph and Briskin, 1989; Kasamo, 1988) and phenylglyoxal (Gildensoph and Briskin, 1989). The presence of essential arginine moieties in enzymes that act on anion substrates is common (Gildensoph and Briskin, 1989 and references therein) and it has been pointed out that the guanidium group of arginine would be well suited as a phosphoryl anion recognition site since it is positively charged and planar; facilitating multiple bonds with phosphoryl groups (Riordin, McElvany, and Borders, 1977). The plasma membrane H^+-ATPase is also inhibited by the histidine modifying reagent, diethylpyrocarbonate (Gildensoph and Briskin, 1990) and by the tyrosine modifying reagent, N-acetylimidazole (Ho and Briskin, 1991). For histidine moieties, a possible role in phosphoenzyme formation was suggested from transient state kinetic studies where the pH-dependence of phosphoenzyme formation appeared similar to the titration curve for an imidazole side chain (Briskin, 1988). Previous work with N-acetylimidazole has also suggested involvement of essential tyrosine moieties in the mechanism of the animal cell Na^+,K^+-ATPase (Argüello and Kaplan, 1990).

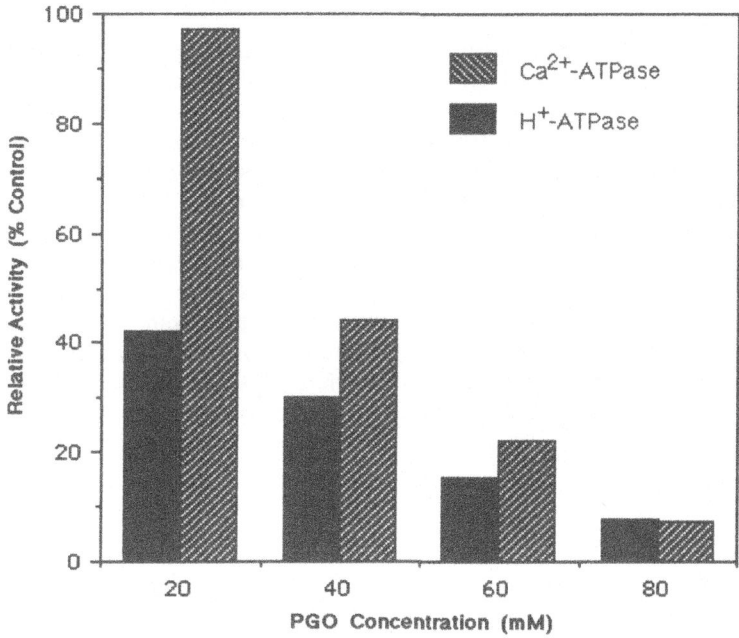

Figure 2 Inhibition of the red beet membrane H^+-ATPase and Ca^{2+}-ATPase by phenylglyoxal (PGO). Red beet plasma membrane fractions were incubated with the indicated concentration of PGO for 10 minutes at 25° C as described by Gildensoph and Briskin (1989). Following incubation with PGO, ATP hydrolytic activity at pH 6.5 (H^+-ATPase) and ATP-dependent $^{45}Ca^{2+}$ uptake (Ca^{2+}-ATPase) were measured. The ATPase assay was conducted as described by Gildensoph and Briskin (1989) while the assay for ATP-dependent $^{45}Ca^{2+}$ uptake was conducted as described by Williams *et al.* (1990).

At present, relatively little is known regarding possible essential amino acid residues involved in the mechanism of the plant plasma membrane Ca^{2+}-ATPase. In preliminary studies, we have characterised the sensitivity of its transport activity to derivatisation with the arginine-modifying reagent phenylglyoxal (Figure 2). As with other P-type transport ATPases (Gildensoph and Briskin, 1989 and references therein), the plant plasma membrane Ca^{2+}-ATPase was inhibited by phenylglyoxal suggesting involvement of essential arginine residues in its mechanism.

However, in the lower range of phenylglyoxal concentrations tested, the plasma membrane Ca^{2+}-ATPase displayed a lesser sensitivity to this modification reagent than the plasma membrane H^+-ATPase. This comparison was based upon both activities being measured in plasma membrane vesicles following derivatisation with this reagent for 10 minutes at room temperature. This difference in the sensitivity to phenylglyoxal may be due to structural differences at the active site for substrate hydrolysis which render arginine moieties less susceptible to attack by this modification reagent.

4.   Differences in Substrate Specificity for Nucleoside Phosphate Compounds

For many transport ATPases, a key characteristic is a high specificity for ATP, as the substrate for ATP hydrolytic or ATP-driven transport activity (Leonard, 1983). This property is displayed by the plant plasma membrane H^+-ATPase and was important for distinguishing its hydrolytic activity from that of non-specific phosphatases in early studies with membrane fractions (Leonard and Hodges, 1980 and references therein). It was then quite surprising when studies indicated that the plasma membrane Ca^{2+}-ATPase could use GTP or ITP as substrates for activity at a level of about 50 to 70% of that observed for ATP (Giannini, Ruiz-Cristin, and Briskin, 1987; Williams *et al.*, 1990; Gräf and Weiler, 1989; Rasi-Caldogno, Pugliarello, Olivari, and De Michelis, 1989). As these measurements were based upon assays of $^{45}Ca^{2+}$ transport activity, the use of other nucleoside phosphate compounds by the plasma membrane Ca^{2+}-ATPase cannot be attributed to non-specific phosphatase activity.

This difference in substrate specificity for the plasma membrane H^+-ATPase and Ca^{2+}-ATPase suggests structural differences between these enzymes at their catalytic active sites. Furthermore, several conclusions can be drawn from a consideration of the chemical structure of the "handle" region of the nucleoside phosphate compounds that can be used by these transport pumps (Figure 3).

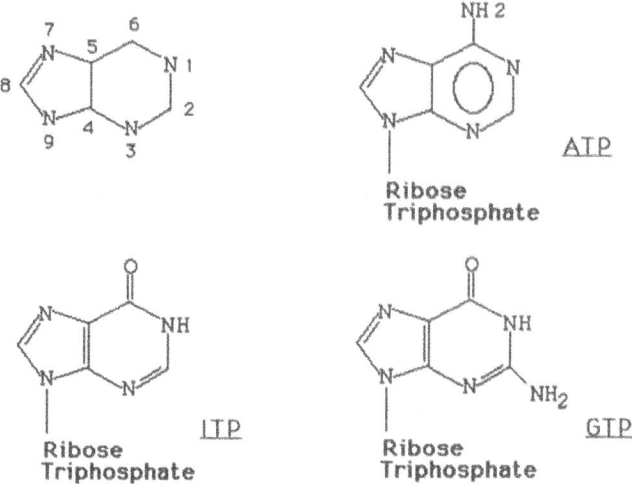

Figure 3   Structure of the purine "handle" region for substrates used by either the plasma membrane H^+-ATPase (ATP) or Ca^{2+}-ATPase (ATP, ITP or GTP).

That only purine compounds can serve as alternative substrates for the plasma membrane Ca^{2+}-ATPase would suggest that the double heterocyclic ring structure is clearly important for this enzyme as well as for the plasma membrane H^+-ATPase. While the amine present at the number 6 position in the purine ring of ATP may be important for substrate recognition by the plasma membrane H^+-ATPase, this is clearly not true for the plasma membrane Ca^{2+}-ATPase since for both ITP and GTP, a carbonyl oxygen is present at this position. However, since activity is reduced from 30 to 50% with these substrates relative to ATP, interaction with the number 6 amine may be more favourable for substrate binding by the plasma membrane Ca^{2+}-ATPase. The observation that similar levels of transport occur for the plasma membrane Ca^{2+}-ATPase using either GTP or ITP would further indicate that amine substitution of the purine ring at the number 2 position has relatively little impact upon activity. That the plasma membrane Ca^{2+}-ATPase can use GTP at all would suggest a possible greater "flexibility" of the active site at this region to accommodate the presence of the amine group.

5.   Differences in Sensitivity to Erythrosin B

Although both the plasma membrane H^+-ATPase and Ca^{2+}-ATPase can be inhibited by erythrosin B, the latter enzyme shows a profound sensitivity to this compound. For both the Ca^{2+} transport (Williams et al., 1990; Rasi-Caldogno et al., 1989) and the ATP hydrolytic (Rasi-Caldogno et al., 1989) activities of the plasma membrane Ca^{2+}-ATPase, substantial inhibition can be achieved at erythrosin B concentrations less then 1 μM. In contrast, erythrosin B concentrations exceeding 100 μM are required to inhibit (>85%) the plasma membrane H^+-ATPase (Coccuci, 1986). In studies with plasma membrane vesicles, erythrosin B concentrations which cause complete inhibition of Ca^{2+} transport by the Ca^{2+}-ATPase have little or no effect upon H^+ transport mediated by the plasma membrane H^+-ATPase (Williams et al., 1990).

This difference in the sensitivity of the plasma membrane H^+-ATPase and Ca^{2+}-ATPase to erythrosin B may again be due to the very same structural differences in active site structure responsible for the differing substrate specificities of these enzymes. This hypothesis is based upon the observation that erythrosin B is an iodinated fluorescein dye and fluorescein derivatives have been shown to inhibit transport ATPases by binding to the region of their active sites responsible for recognition of the adenine "handle" portion of ATP (Tonomura, 1986). This has formed the basis for use of FITC to derivatise a specific lysine group in several P-type ATPases involved in this recognition process (see previous discussion). Hence, the greater "flexibility" of the plasma membrane Ca^{2+}-ATPase active site (relative to the H^+-ATPase) may allow it to better accommodate the iodinated groups on erythrosin B and dramatically increase the sensitivity of this enzyme to this compound.

We have conducted studies with erythrosin isothiocyanate (EITC) as a preliminary step to actually determine whether erythrosin B is acting at the active site of the plasma membrane Ca^{2+}-ATPase in a manner similar to other fluorescein derivatives. This compound would be similar to FITC in having a reactive isothiocyanate group that could form a covalent bond with lysine residues (Means and Feeney, 1974). Treatment of the plasma membrane Ca^{2+}-ATPase in vesicles with EITC leads to an irreversible inhibition of ATP-dependent Ca^{2+} transport activity (Briskin, Gildensoph, and Basu, 1990). This inhibitory chemical modification by EITC appears specific for the plasma membrane Ca^{2+}-ATPase since ATP-dependent H^+ transport mediated by the plasma membrane H^+-ATPase is unaffected by treatment with this reagent (Figure 4). Studies are currently in progress to develop a radiolabelled version of EITC useful for labelling and subsequent identification of the derivatised lysine residue associated with the plasma membrane Ca^{2+}-ATPase.

6. Role of Counterions in the Transport Reaction Mechanism

Although research on the plant plasma membrane H^+-ATPase and Ca^{2+}-ATPase has focused primarily on their respective roles in H^+ or Ca^{2+} extrusion, there has also been interest in determining whether these enzymes might be involved in the primary transport of other ions as well. Such additional primary roles in ion transport would have important physiological and biophysical implications for the functioning of these transport pumps.

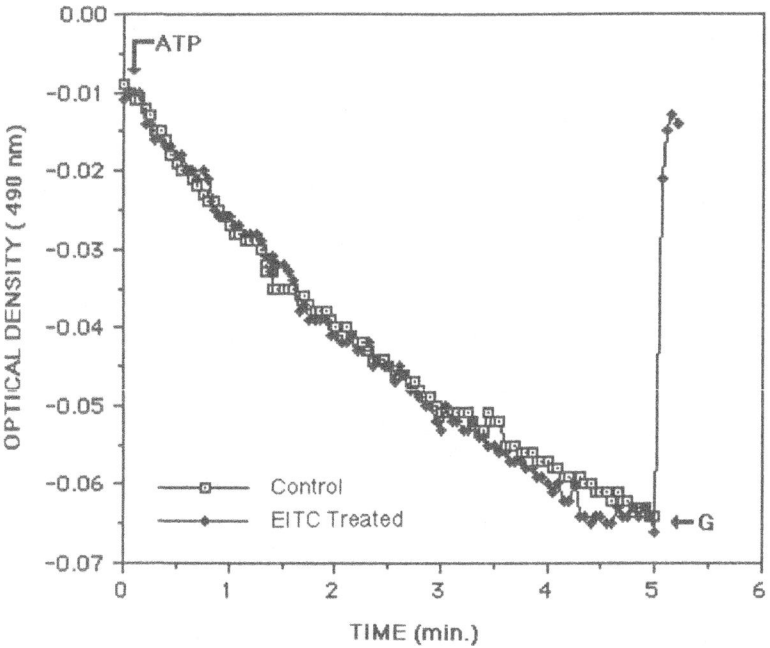

Figure 4 Effect of erythrosin isothiocyanate (EITC) on ATP-dependent H^+ transport by the plasma membrane H^+-ATPase. Red beet plasma membrane vesicles were treated with 200 nM EITC for 15 minutes at 25° C. Aliquots were withdrawn and then assayed for ATP-dependent H^+ transport in the presence of 250 mM sorbitol, 3 mM ATP, 3 mM $MgSO_4$, 25 mM BTP/Mes pH 6.5, 50 mM KCl and 5 μM acridine orange. Transport was initiated by the addition of ATP and the optical density change at 490 nm representing H^+ transport into the vesicles was monitored. Where indicated (G), 5 μM gramicidin D was added to collapse the gradient and demonstrate the presence of a ΔpH. The control sample represents plasma membrane vesicles incubated under the same conditions as the EITC treated-samples except that EITC was not present.

In early studies on the plasma membrane H^+-ATPase, it was suggested that this enzyme might also be involved in K^+ uptake and thus act as an ATP-dependent H^+/K^+ exchange pump (Briskin, 1990a and references therein). Such a role for the plant plasma membrane H^+-ATPase would be analogous to the animal cell Na^+,K^+-ATPase where Na^+ extrusion generates a sodium electrochemical gradient utilised to drive other transport processes and K^+ uptake would be directly mediated by this protein (Tonomura, 1986 and references therein).

This concept was based on the initial observation that relative rates of K^+ influx into root tissue from several species (maize, wheat, oats, barley) appeared to correlate with the level of K^+-stimulated ATPase activity associated with microsomal membrane fractions prepared from these roots (Fisher and Hodges, 1969; Fisher, Hansen, and Hodges, 1970). Furthermore, K^+ uptake often displays a complex kinetic profile similar to that observed for K^+-simulation of ATP hydrolytic activity (Leonard and Hodges, 1973). Potassium stimulation of ATP hydrolysis is observed typically for those enzymes (e.g. animal Na^+-ATPase, gastric H^+,K^+-ATPase) that transport this cation directly. This stimulation arises because K^+ serves to increase the rate of phosphoenzyme turnover in their reaction mechanisms (Skou, 1990; Rabon and Reubon, 1990). From mechanistic studies conducted on the plant plasma membrane H^+-ATPase, it was also shown that K^+ is involved in increasing the rate of phosphoenzyme turnover (Briskin, 1990a and references therein). On the other hand, the small magnitude of these K^+ effects (Briskin, 1990a and references therein), other inconsistencies between the plant H^+-ATPase and K^+-transporting ATPases (Serrano, 1984) and the observation that H^+ efflux can occur independently of K^+ uptake (Kochian, Shaff, and Lucas, 1989) has led many workers to question the concept of a H^+/K^+-ATPase. Hence, a current viewpoint is that the plasma membrane H^+-ATPase acts only as a primary H^+ extrusion pump and K^+ uptake would likely occur by mechanisms involving either secondary transport ($\Delta\psi$-driven K^+ channels or K^+ uniports, H^+/K^+ symports) or a separate primary K^+ transporter (Kochian et al., 1989).

In their attempts to determine whether Ca^{2+} efflux at the plant plasma membrane was mediated by a separate Ca^{2+}-ATPase or by an nH^+/Ca^{2+} antiport, Rasi-Caldogno, Pugliarello and DeMichelis (1987) found evidence for an association of H^+ counterflux with Ca^{2+} transport mediated by the plasma membrane Ca^{2+}-ATPase. While ATP-driven Ca^{2+} transport in their studies with plasma membrane vesicles was insensitive to protonophores, effects of Ca^{2+} upon ΔpH and $\Delta\psi$ were observed that were inhibited by low concentrations of erythrosin B. These workers concluded that a plasma membrane Ca^{2+}-ATPase was present in the vesicles that mediated nH^+/Ca^{2+} exchange. This would be similar to what is observed for the plasma membrane Ca^{2+}-ATPase associated with animal cells (Schatzman, 1989). However, a significant problem in directly examining an nH^+/Ca^{2+}-ATPase in isolated plasma membrane vesicles is the large background of ATP-driven H^+ ATPase.

The plasma membrane Ca^{2+}-ATPase can use alternative substrates such as ITP and GTP and this could provide a means for examining H^+/Ca^{2+} exchange activity associated with this enzyme in isolated vesicles where the plasma membrane H^+-ATPase is also present. Because the plasma membrane H^+-ATPase is highly specific for ATP as the substrate for driving transport (Briskin, 1990a), H^+ flux that would be driven by these alternative substrates might reflect H^+/Ca^{2+} exchange activity mediated by the plasma membrane Ca^{2+}-ATPase. However, a significant problem in measuring this H^+ transport activity is that it would involve H^+ efflux from the vesicles but the optical probes most frequently used in plant vesicle transport studies involving H^+ flux (e.g. weak base probes) sense acid-interior pH gradients (ΔpH) generated by H^+ influx (Briskin, 1990c and references therein). In our present studies, we are examining the possibility of using water soluble pH indicator type dyes for measuring the internal alkalinization of plasma membrane vesicles that might be associated with H^+/Ca^{2+} exchange. The use of these dyes would necessitate their entrapment inside the vesicles where they would serve to sense any internal scalar pH changes associated with H^+ efflux driven by an alternative substrate.

7.   Perspective

Although the general mechanistic features of the plasma membrane H^+-ATPase and Ca^{2+}-ATPase are sufficiently similar to allow both proteins to be classified as P-type ATPases, these enzymes show distinct properties most likely related to structural differences associated with their catalytic active sites. This will perhaps become more evident as molecular characterisation of the enzymes progresses even further. The ability of the plasma membrane Ca^{2+}-ATPase to use alternative substrates (GTP or ITP) has proven useful for considering possible differences in substrate determinants recognised by these enzymes and could be useful for examining whether H^+/Ca^{2+} exchange is associated with this enzyme. The presence of a Ca^{2+}-ATPase at the plasma membrane in addition to the H^+-ATPase clearly provides a pathway for active Ca^{2+} efflux from the cytoplasm that can operate independently of the proton electrochemical established by the proton pump. With current research suggesting that the Ca^{2+} regulatory system might be involved in modulating the plasma membrane H^+-ATPase (Sussman, Harper, DeWitt, Schaller, and Sheaham, 1991), this independence may be important for insuring restoration of Ca^{2+} gradients to pre-stimulus levels during signal transduction events that might affect the activity of the H^+-pump.

REFERENCES

ARGÜELLO, J.M. and KAPLAN, J.H., 1990. N-acetylimidazole inactivates renal Na,K-ATPase by disrupting ATP binding to the catalytic site. *Biochemistry*, **29**, 5775-5785.

BRIARS, S.A. and EVANS, D.H., 1989. The calmodulin-simulated ATPase of maize coleoptiles forms a phosphorylated intermediate. *Biochemical and Biophysical Research Communications*, **159**, 185-91.

BRISKIN, D.P., 1986. Intermediate reaction states of the red beet plasma membrane ATPase. *Archives of Biochemistry and Biophysics*, **248**, 106-15.

BRISKIN, D.P., 1988. Phosphorylation and dephosphorylation reactions of the red beet plasma membrane ATPase studies in the transient state. *Plant Physiology*, **88**, 84-91.

BRISKIN, D.P., 1990a. The plasma membrane H^+-ATPase of higher plant cells:: biochemistry and transport function. *Biochimica et Biophysica Acta*, **1019**, 95-109.

BRISKIN, D.P., 1990b. Ca^{2+}-translocating ATPase of the plant plasma membrane. *Plant Physiology*, **94**, 397-400.

BRISKIN, D.P., 1990c. Transport in plasma membrane vesicles - approaches and perspectives. In *The Plant Plasma Membrane*. Eds C. Larsson, I.M. Møller. Springer-Verlag, Berlin. pp 154-181.

BRISKIN, D.P. and POOLE, R.J. 1983a. Evidence for a β-aspartyl phosphate residue in the phosphorylated intermediate of the red beet plasma membrane ATPase. *Plant Physiology*, **72**, 1133-1135.

BRISKIN, D.P. and POOLE, R.J., 1983b. Role of magnesium in the plasma membrane ATPase of red beet. *Plant Physiology*, **71**, 969-971.

BRISKIN, D.P., GILDENSOPH, L.H. and BASU, S., 1990. Characterisation of the Ca^{2+}-transporting ATPase of the plant plasma membrane using isolated membrane vesicles. In *Calcium in Plant Growth and Development*. Eds R.T. Leonard, P.K. Hepler. American Society of Plant Physiologists. pp 46-54.

CANTLEY, L.C., CARILLI, C.T., FARLEY, R.A. and PERLMAN, D.M., 1982. Location of binding sites on the (Na,K)-ATPase for fluorescein-5'-isothiocyanate and ouabain. *Annals of the New York Academy of Sciences*, **402**, 289-291.

CARAFOLI, E., 1991. The calcium pumping ATPase of the plasma membrane. *Annual Review of Physiology*, **53**, 531-547.

CHASTEEN, N.D., 1983. The biochemistry of vanadium. *Structure and Bonding*, **53**, 107-138.

COCCUCI, M.C., 1986. Inhibition of plasma membrane and tonoplast ATPases by erythrosin B. *Plant Science*, **47**, 21-27.

DEPONT, J.J.H.M., SCHOOT, B.M., VANPROOIGEN-VANEEDEN, A. and BONTING, S.L., 1977. An essential arginine residue in the ATP-binding centre of (Na$^+$ + K$^+$)-ATPase. *Biochimica et Biophysica Acta*, **482**, 213-227.

EYZRAGUIRRE, J., 1987. Chemical modification of enzymes - an overview. The use of group-specific reagents. In *Chemical Modification of Enzymes:: Active Site Studies*. Ed J. Eyzraguirre, J. Wiley and Sons, New York. pp 3-15.

FARLEY, R.A., TRAN, C.M., CARILLI, C.T., HAWKE, D. and SHIVELY, J.E., 1984. The amino acid sequence of a fluorescein-labelled peptide from the active site of (Na$^+$,K$^+$)-ATPase. *Journal of Biological Chemistry*, **259**, 9532-9535.

FISHER, J. and HODGES, T.K., 1969. Monovalent ion stimulated adenosine triphosphatase from oat. *Plant Physiology*, **44**, 385-395.

FISHER, J.D. HANSEN, D., and HODGES, T.K., 1970. Correlation between ion fluxes and ion-stimulated adenosine triphosphatase activity of plant roots. *Plant Physiology*, **46**, 812-814.

FUTAI, M., NOUMI, T. and MAEDA, M., 1989. ATPase synthase (H^+-ATPase): Results by combined biochemical and molecular biological approaches. *Annual Review of Biochemistry*, **58**, 111-136.

GIANNINI, J.L., RUIZ-CRISTIN, J.L., and BRISKIN, D.P., 1987. Calcium transport in sealed vesicles from red beet (*Beta vulgaris* L.) storage tissue. II. Characterisation of $^{45}Ca^{2+}$ uptake into plasma membrane vesicles. *Plant Physiology*, **85**, 1137-1142.

GILDENSOPH, L.H. 1989. Chemical modification of essential amino acid moieties associated with the red beet (*Beta vulgaris* L.) plasma membrane H^+-ATPase. Ph.D. thesis, University of Illinois.

GILDENSOPH, L.H., and BRISKIN, D.P., 1989. Modification of an essential arginine residue associated with the plasma membrane ATPase of red beet (*Beta vulgaris* L.) storage tissue. *Archives of Biochemistry and Biophysics*, **271**, 254-59.

GILDENSOPH, L.H., and BRISKIN, D.P., 1990. Modification of the red beet plasma membrane H^+-ATPase by diethylpyrocarbonate. *Plant Physiology*, **94**, 696-703.

GRÄF, P. and WEILER, E.W., 1989. ATP-driven Ca^{2+} transport in sealed plasma membrane vesicles prepared by aqueous polymer two-phase partitioning from leaves of *Commelina communis*. *Physiologia Plantarum*, **75**, 469-78.

HO, I.,and BRISKIN, D.P., 1991. Characterisation of essential tyrosine residues associated with the plasma membrane H^+-ATPase of red beet storage tissue. *Plant Physiology Supplement*, **96**, #933a.

KARLISH, S.J.D., 1989. The mechanism of cation transport by the Na^+,K^+-ATPase. In *Ion Transport*. Eds K. Keeling, C. Benham. Academic Press, London. pp 19-34.

KASAI, M. and MUTO, S., 1990. Ca^{2+} pump and Ca^2/H^+ antiporter in plasma membrane vesicles isolated by aqueous two-phase partitioning from corn leaves. *Journal of Membrane Biology*, **114**, 133-142.

KASAMO, K., 1988. Essential arginyl residues in the plasma membrane H^+-ATPase from *Vigna radiata* L. (mung bean) roots. *Plant Physiology*, **87**, 126-129.

KASHER, J.S., ALLEN, K.E., KASAMO, K. and SLAYMAN, S.W., 1986. Characterisation of an essential arginine residue in the plasma membrane H^+-ATPase of *Neuorspora crassa*. *Journal of Biological Chemistry*, **261**, 10808-10813.

KNOWLES, J.R., 1980. Enzyme-catalysed phosphoryl transfer reactions. *Annual Review of Biochemistry*, **49**, 877-919.

KOCHIAN, L.V., SHAFF, J.E., and LUCAS, W.J., 1989. High affinity K^+ uptake in maze roots. A lack of coupling with H^+ efflux. *Plant Physiology*, **91**, 1202-1211.

LEONARD, R.T., 1983. Potassium transport and the plasma membrane ATPase in plants. In *Metals and Micronutrients: Uptake and Utilisation by Plants*. Eds D.A. Robb, W.S. Pierpoint. Academic Press, London. pp 71-86.

LEONARD, R.T., and HODGES, T.K., 1973. Characterisation of plasma membrane-associated adenosine triphosphatase activity of oat roots. *Plant Physiology*, **52**, 6-12.

LEONARD, R.T., and HODGES, T.K., 1980. The plasma membrane. In *The Biochemistry of Plants. A Comprehensive Treatise*. Eds P.K. Stumpf, E.E. Conn. Academic Press, New York. Volume 1, pp 163-181.

MACARA, I.G. 1980. Vanadium - an element in search of a role. *Trends in Biochemical Sciences*, **5**, 92-94.

MALATIALY, L., GREPPIN, H. and PENEL, C., 1988. Calcium uptake by tonoplast and plasma membrane vesicles from spinach leaves. *FEBS Letters*, **223**, 196-200.

MARMÉ, D., 1989: The role of calcium and calmodulin in signal transduction. In *Second Messengers in Plant Growth and Development*. Eds W.F. Boss, D.J. Morré, Alan R. Liss, Inc. pp 57-80.

MEANS, G.E. and FEENEY, R.E., 1974. Chemical Modification of Proteins. Holden-Day, Inc., San Francisco. 254 pages.

MURPHY, A.J., 1976. Arginyl residue modification of the sarcoplasmic reticulum ATPase protein. *Biochemical and Biophysical Research Communications*, **70**, 1048-1054.

PARDO, J.P., and SLAYMAN, C.W., 1988. The fluorescein isothiocyanate-binding site of the plasma membrane H^+-ATPase of *Neurospora crassa*. *Journal of Biological Chemistry*, **263**, 19664-18668.

PEDERSON, P.L., and CARAFOLI, E., 1987. Ion motive ATPases. I. Ubiquity, properties and significance to cell function. *Trends in Biochemical Sciences*, **12**, 146-150.

PICK, U., and KARLISH, S.J.D., 1980. Indications for oligomeric structure and for conformational changes in sarcoplasmic reticulum Ca^{2+}-ATPase labelled selectively with fluorescein. *Biochimica et Biophysica Acta*, **626**, 255-61.

23

RABON, E.C. and REUBEN, M.A., 1990. The mechanism and structure of the gastric H,K-ATPase. *Annual Review of Physiology*, **52**, 321-344.

RASI-CALDOGNO, F., PUGLIARELLO, M.C. and DEMICHELIS, M.I., 1987. The Ca^{2+}-transport ATPase of plant plasma membrane catalyses a nH^+/Ca^{2+} exchange. *Plant Physiology*, **83**, 994-1000.

RASI-CALDOGNO, F., PUGLIARELLO, M.C., OLIVARI, C., and DEMICHELIS, M.I., 1989. Identification and characterisation of the Ca^{2+}-ATPase which drives active transport of Ca^{2+} at the plasma membrane of radish seedlings. *Plant Physiology*, **90**, 1429-1434.

RIORDIN, J.F., McELVANY, K.D., and BORDERS, Jr., C.L., 1977. Arginyl residues: Anion recognition sites in enzymes. *Science*, **195**, 884-886.

RUDOLF, F.B., 1983. Product inhibition and abortive complex formation. In *Contempary Enzyme Kinetics and Mechanism*. Ed D.L. Purich. Academic Press, New York. pp 207-232.

SANDERS, D., and SLAYMAN, C., 1989. Transport at the plasma membrane of plant cells: a review. In *Plant Membrane Transport, the Current Position*. Eds J. Dainty, M.I. DeMichelis, E. Marrè, and F. Rasi-Caldogno, Elsevier, Amsterdam. pp 3-11.

SCHATZMANN, H.J., 1989: The calcium pump of the surface membrane and of the sarcoplasmic reticulum. *Annual Review of Physiology*, **51**, 473-485.

SERRANO, R., 1984. Plasma membrane ATPase of fungi and plants as a novel type of proton pump. *Current Topics in Cell Regulation*, **23**, 87-126.

SKOU, J.C., 1990. The energy coupled exchange of Na^+ for K^+ across the cell membrane. The Na^+,K^+-pump. *FEBS Letters*, **268**, 314-324.

STONE, D.K., CRIDER, B.P., SÜDHOF, T.C. and XIE, X-S., 1990. Vacuolar proton pumps. *Journal of Bioenergetics and Biomembranes*, **21**, 605-620.

SUSSMAN, M.R., HARPER, J.F., DEWITT, N., SCHALLER, E., and SHEAHAM, J., 1991. Plasma membrane cation pumps in *A. thaliana*: number of isoforms, expression and regulation of activity. *Plant Physiology Supplement*, **96**, #1.

TONOMURA, Y., 1986: Energy Transducing ATPases-Structure and Kinetics. Cambridge University Press, Cambridge.

WALDERHAUG, M.O., POST, R.L., SACCOMANI, G., LEONARD, R.T., and BRISKIN, D.P., 1985. Structural relatedness of three ion-transport adenosine triphosphatases around their active sites of phosphorylation. *Journal of Biological Chemistry*, **260**, 3852-59.

WILLIAMS, L.E., SCHUELER, S.B., and BRISKIN, D.P., 1990. Further characterisation of the red beet plasma membrane Ca^{2+}-ATPase using GTP as an alternative substrate. *Plant Physiology*, **92**, 747-754.

Vacuolar H$^+$-translocating Inorganic Pyrophosphatase: Biochemistry and Molecular Biology

Philip A. Rea[1], Yongcheol Kim[1], Vahe Sarafian[1,2], Ronald J. Poole[2] and Christopher J. Britten[1]

1. Plant Science Institute
    Department of Biology
    University of Pennsylvania
    Philadelphia, USA

2. Department of Biology
    McGill University
    Montreal, Canada

Abbreviations

$\Delta\bar{\mu}_{H+}$, H$^+$-electrochemical potential difference; AVP, cDNA insert encoding *Arabidopsis* Vacuolar (H$^+$-translocating) Pyrophosphatase; NEM, N-ethylmaleimide; pAVP, plasmid construct containing AVP cDNA insert; PP$_i$, inorganic pyrophosphate (diphosphate); V-PPase, vacuolar H$^+$-PPase.

1.   Introduction

General acceptance of the chemiosmotic hypothesis (Mitchell 1961) has given rise to the view that membrane-bound H$^+$-pumps constitute the primary transducers by means of which living cells inter-convert light, chemical and electrical energy. Through the establishment and maintenance of trans-membrane electrochemical gradients, H$^+$ pumps energise the transport of other solutes or, in the special case of the energy-coupling membranes of mitochondria, chloroplasts and bacteria, transduce the H$^+$-electrochemical gradient generated by membrane-linked anisotropic redox reactions to the synthesis of ATP (Harold, 1986). Given the multitude of biological reactions energised by ATP, primary H$^+$-translocation and the inter-conversions of ATP have come to be recognised as the principal generators of usable energy in the cell. Against this background it is, therefore, surprising to find that the vacuolar

Transport and Receptor Proteins of Plant Membranes
Edited by D.T. Cooke and D.T. Clarkson, Plenum Press, New York, 1992

25

membrane (tonoplast) of plant cells contains not only a V-("vacuolar") type H$^+$-ATPase (EC 3.6.1.3) (Rea and Sanders, 1987; Nelson and Taiz, 1989) but also an inorganic pyrophosphate- (PP$_i$) energised H$^+$ pump (V-type H$^+$-PPase; EC 3.6.1.1) (Rea and Sanders, 1987). Both enzymes catalyse inward electrogenic H$^+$-translocation (from cytosol to vacuole lumen) but the vacuolar-("V-type") H$^+$-PPase has the unusual characteristic of exclusively utilising PP$_i$ as energy source (Rea and Poole, 1986).

2.   Biochemistry

2.1   Initial characterisation

The notion that plants contain a membrane-associated PPase is not new. Karlsson in the mid-seventies (Karlsson, 1975) demonstrated a K$^+$-stimulated PPase activity in mixed membrane fractions from sugar-beet (*Beta vulgaris*) seedlings and Walker and Leigh (1981) characterised an Mg^{2+}-dependent PPase with qualitatively similar kinetics associated with isolated vacuoles. However, it was several years before the transport function of the enzyme was delineated. It had been shown that PP$_i$ will support H$^+$-translocation by tonoplast vesicles isolated from plant tissues, (Churchill and Sze, 1983; Bennett, O'Neill, and Spanswick, 1984) but it was not known at the time if the observed PP$_i$-dependent intravesicular acidification resulted from the activity of the PPase described by Karlsson (1975) and Walker and Leigh (1981), or whether the, then recently defined tonoplast H$^+$-ATPase was capable of utilising PP$_i$ as an energy source under some circumstances.

The initial indication that PP$_i$-dependent H$^+$-translocation is mediated by the V-PPase and not the V-ATPase came from a direct comparison of the substrate, mineral ion and effector-sensitivities of PP$_i$ hydrolysis and PP$_i$-dependent H$^+$-translocation (Rea and Poole, 1985). Briefly, ATP hydrolysis and PP$_i$ hydrolysis by uncoupled tonoplast vesicles do not interact; PP$_i$ hydrolysis and PP$_i$-dependent H$^+$-translocation are maximally stimulated by K$^+$ whereas the V-type H$^+$-ATPase is selectively activated by halides; PP$_i$ hydrolysis, unlike ATP hydrolysis, is not subject to inhibition by nitrate. Subsequent chromatographic resolution of the V-PPase and V-ATPase (Rea and Poole, 1986; Wang, Leigh, Kaestner and Sze, 1986) confirmed the conclusions drawn from the kinetic studies by showing that the separated V-PPase is specific for MgPP$_i$ (or Mg$_2$PP$_i$) as substrate while the separated V-ATPase shows essentially no activity towards this compound (Rea and Poole, 1986).

2.2   Species distribution

The V-type H$^+$-PPase appears to be ubiquitous in the vacuolar membranes of higher plant cells. PP$_i$-dependent H$^+$-translocation, K$^+$-stimulated and/or molybdate-insensitive PP$_i$ hydrolysis have been demonstrated in the vacuolar membranes of monocotyledons, dicotyledons, C3, C4 and CAM plants, leaves, roots and Characean algae (e.g. Rea and Sanders, 1987). Less clear is the situation in *non-plant* vacuolate cells, particularly yeast. On the one hand, Kulakovskaya, Lichko, and Okorokov (1989), provide evidence for MgPP$_i$ hydrolysis and MgPP$_i$-dependent H$^+$-translocation by vacuolar membrane vesicles isolated from *Saccharomyces carlsbergensis*. On the other hand, vacuolar membranes isolated from *Saccharomyces cerevisiae* lack polypeptides with immunological cross-reactivity towards the V-PPase from plants (Kim, E.J., and Rea, P.A., unpublished) and genomic Southern analyses of *S. cerevisiae*, *S. carlsbergensis* and *Schizosaccharomyces pombe* DNA using cDNAs encoding the V-PPase from *Arabidopsis* as probes (see *Section II*) fail to show homologues in the yeast genome (Eisman, R., Kim, E.J., and Rea, P.A., unpublished). If yeasts contain a vacuolar H^+-PPase it is apparently divergent from the plant enzyme. Alternatively, the activity measured by Kulakovskaya *et al.*, (1989) corresponds to contaminating mitochondrial H^+-PPase, which does not show homology with the V-PPase (*Section 3.4*). While it remains to be determined if animal cells contain endomembrane V-type H^+-PPases, preliminary experiments (Rea, P.A., and Apps, D.K., unpublished) indicate that chromaffin granule ghosts - which are known to possess a V-type H^+-ATPase that is immunologically cross-reactive with (Manolson, Percy, Apps, Xie, Stone, Harrison, Clark, and Poole, 1987) and homologous to the plant V-type H^+-ATPase (Nelson and Taiz, 1989) - isolated from the adrenal medulla, are not capable of PP_i-dependent H^+-translocation. [For considerations of the presence of "V-type" H^+-PPases in phototrophic bacteria see *Section 3.4*].

2.3   Bio-energetic impact

The V-PPase is a major vacuolar membrane component of plants. The enzyme is capable of establishing and maintaining a $\Delta \bar{\mu}_{H+}$ of similar, or greater, magnitude than the V-ATPase on the same membrane; the extreme case being tonoplast vesicles isolated from etiolated hypocotyls of mung bean (*Vigna radiata*) in which the V-PPase is 4-6 times more active that the V-ATPase, whether activity is measured as H^+-translocation or substrate hydrolysis (Rea, Britten, and Sarafian, 1992). The specific activity of the purified enzyme (12-20 μmol/mg.min; Britten, Turner, and Rea, 1989; Maeshima and Yoshida, 1989; Rea *et al.*, 1992) is consistent with a turnover number of 30-60 s^{-1} and abundance estimates for the M_r 64,500-73,000, substrate-binding subunit of the V-PPase (Britten *et al.*, 1989; *Section 2.4*) yield values of between 1% (Britten *et al.*, 1989) and 5-10%, respectively (Maeshima and Yoshida, 1989; Rea *et al.*, 1992). Thus, the potential bioenergetic impact of the V-PPase is great, especially when account is taken of the fact that the vacuole can occupy 90-99% of the total intracellular volume of a mature plant cell.

2.4   Physical identity

There have been conflicting reports in the literature concerning the polypeptide composition of the V-PPase. The major subunit of the enzyme has been attributed to polypeptides of M_r 64,500 (*Beta vulgaris*, Britten *et al.*, 1989), M_r 67,000 (*Beta vulgaris*, Britten *et al.*, 1989), M_r 67,000 (*Beta vulgaris*, Sarafian and Poole, 1989), M_r 73,000 (*Vigna radiata*, Maeshima and Yoshida, 1989) and M_r 37,000-44,000 (*Zea mays*, Chanson and Pilet, 1989; and *Acer pseudoplatanus*, Fraichard, Magnin, and Pugin, 1991). However, through the combined application of independent purification protocols, affinity-labelling, sequencing and immunological techniques it is now clear that the major polypeptides associated with at least four of these preparations and the H^+-PPase of *Arabidopsis*, correspond to the same moiety (Rea *et al.*, 1992). First, the M_r 64,500 polypeptide purified from *Beta* co-migrates with the major polypeptide of the enzyme from *Vigna*, purified by an independent procedure, when the samples are subjected to SDS-PAGE under identical conditions and the two polypeptides show indistinguishable patterns of ligand-modified labelling by [^{14}C]-N-ethylmaleimide (NEM) (*Section 2.5*). Second, the M_r 64,500 and M_r 67,000 polypeptides isolated from *Beta* by independent procedures (cf. Britten *et al.*, 1989 *versus* Sarafian and Poole, 1989) co-migrate when electrophoresed under the same conditions and yield tryptic fragments with sequences identical to each other (Rea *et al.*, 1992) and to the deduced translation product of H^+-PPase clones isolated by antibody and hybridisation screens (Sarafian, Kim, Poole, and Rea, 1992). Third, immunoblots of membranes prepared from *Arabidopsis, Beta, Vigna* and

Zea and probed with antibody affinity-purified against the M_r 66,000 polypeptide of *Vigna* yield a single band migrating at M_r 64,500-66,800 in all four preparations (Rea *et al.*, 1992).

2.5 Catalytic function of major subunit

Direct participation of the M_r 64,500-66,800 polypeptide in substrate-binding is implied by its kinetics of labelling with [^{14}C]-NEM (Britten *et al.*, 1989; Rea *et al.*, 1992). The V-PPase activities of both *Beta* and *Vigna* are subject to ligand-modified irreversible inhibition by NEM (Figure 1A). Inhibition is pseudo-first order and approximates the relationship:

$$\%C = 100\ (1 - \exp)(k^o t[\text{NEM}])$$

where %C is percentage inhibition (% control), k^o is the first order rate constant, t is time and [NEM] is the concentration of NEM.

Quantitative protection is conferred by MgPP$_i$:

$$(k^o_{beta} = 0.22 \times 10^3\ \text{M}^{-1}\ \text{min}^{-1};\quad k^o_{Vigna} = 0.25 \times 10^3\ \text{M}^{-1}\ \text{min}^{-1})$$

whereas free PP$_i$ increases the potency of NEM by a factor of two:

$$(k^o_{Beta} = 4.09 \times 10^3\ \text{M}^{-1}\ \text{min}^{-1};\quad k^o_{Vigna} = 6.51 \times 10^3\ \text{M}^{-1}\ \text{min}^{-1})$$

versus control samples incubated with NEM in the absence of ligands:

$$(k^o_{Beta} = 2.30 \times 10^3\ \text{M}^{-1}\ \text{min}^{-1};\quad k^o_{Vigna} = 3.12 \times 10^3\ \text{M}^{-1}\ \text{min}^{-1}).$$

Inhibition by NEM is independent of membrane protein concentration (Figure 1A, *Inset*) and neither MgPP$_i$ or free PP$_i$ affect the kinetics of inhibition of the tonoplast H$^+$PPase (Britten *et al.*, 1989).

SDS-PAGE and fluorography of tonoplast vesicles labelled with 50 μM[^{14}C]-NEM after pre-treatment with 50 μM[^{14}C]-NEM in the presence of MgPP$_i$ - to block non-protectable, non-essential NEM-reactive groups and protect protectable groups - generates two prominent ^{14}C-labelled polypeptides of M_r 64,500 and 23,000 in *Beta* and one labelled polypeptide of M_r 66,000 in *Vigna* (Figure 1B). In both preparations, the M_r 64,500-66,000 polypeptide, alone, is subject to labelling under conditions which cause maximal inhibition of the H$^+$-PPase: labelling is abolished by MgPP$_i$ and potentiated by free PP$_i$. Thus, the purification and affinity-labelling data are in close agreement. Purified preparations of the V-PPase from *Beta* and *Vigna* are primarily constituted from polypeptides of M_r 64,500 and 66,000, respectively, and the same polypeptides are susceptible to MgPP$_i$-protectable, free PP$_i$-potentiated labelling by [^{14}C]-NEM in isolated tonoplast vesicles. The M_r 64,500 and M_r 66,000 polypeptides are tentatively identified as the catalytic subunit of the respective V-PPase preparations; a conclusion consistent with the results of kinetic investigations which show that MgPP$_i$ (White, Marshall and Smith, 1990), or Mg$_2$PP$_i$ (R A Leigh, unpublished), is the active substrate species.

The M_r 23,000, ^{14}C-labelled polypeptide of *Beta* is unlikely to be a subunit of the H$^+$-PPase. It is absent from *Vigna*, which contains at least 4-fold more H$^+$-PPase activity and is resolved from PPase activity (and the M_r 64,500 subunit) when solubilised membranes are fractionated by FPLC on Mono-Q (Britten *et al.*, 1989). It is maximally labelled by [^{14}C]-NEM when inhibition of the H^+-PPase is minimal, i.e. when $MgPP_i$ is included in the NEM reaction medium (Figure 1A).

3. Molecular Biology

cDNAs encoding the V-PPase of *Arabidopsis thaliana* have recently been cloned and sequenced (Sarafian *et al.*, 1992). The open reading frame of the near-full length clone pAVP-3 (where AVP = *Arabidopsis* Vacuolar Pyrophosphatase) encodes a 770 amino acid polypeptide with a pI of approximately 4.95 and a predicted molecular weight of 80,800 daltons ("81 kDa"). The estimated molecular weight of the protein encoded by the insert is approximately 8 kDa larger than the highest apparent M_r (73,000) determined by SDS-PAGE of the enzyme from *Vigna* (Maeshima and Yoshida, 1989). It is also 13.8-16.3 kDa greater than the estimated M_r (64,500-67,000) of the corresponding subunit from *Beta* (Britten *et al.*, 1989; Sarafian and Poole, 1989) and about 14 kDa greater than the M_r 66,800 immunoreactive polypeptide reported for *Arabidopsis* (Sarafian *et al.*, 1992). This anomalous migration of hydrophobic membrane proteins on SDS-gels is not unusual, because binding of non-saturating amounts of SDS, resulting in the exposure of charged amino acid residues or irregularities in the shape of the SDS-protein complex, can cause large shifts in apparent M_r.

3.1 Identity of cDNA clones

Immunological and direct sequence data confirm the identity and deduced sequence of the clones. The initial clones were isolated by immuno-screens of a λZAP expression library. However, precise alignment was found between the deduced amino acid sequence of the protein encoded by AVP-3, direct internal sequence data acquired from the M_r 64,500-67,000 $MgPP_i$-binding subunit of *Beta* (Britten *et al.*, 1989; Sarafian and Poole, 1989; Rea *et al.*, 1992) and N-terminal sequence data obtained from the corresponding polypeptide purified from *Vigna* (Maeshima and Yoshida, 1989).

In situ tryptic digestion (Aebersold, Leavitt, Saavedra, Hood, and Kent, 1987) of the "M_r 67,000" (*Beta-1*) and "M_r 64,500" (*Beta-2*) substrate-binding subunits of the V-PPase, purified by the methods of Sarafian and Poole (1989) and Britten *et al.*, (1989), respectively, generates peptide fragments with identical sequences. Of a total of 6 tryptic fragments subjected to gas-phase sequence analysis (three from *Beta-1* and three from *Beta-2*), two fragments from each preparation overlap exactly with two of the fragments from the other preparation. Moreover, all six microsequences, including those unique to the two *Beta* V-PPase preparations, are present in the open reading frame of clone pAVP-3 at positions 255, 530, 567 and 722, respectively, in the deduced amino acid sequence. The direct sequence from the M_r 64,500-67,000 subunit of *Beta* and the amino acid sequence deduced from the nucleotide sequence of pAVP-3 show complete identity over a total span of 66 amino acid residues, except for two conservative (Val → Ile, Gln → His) substitutions at positions 266 and 570, respectively, and one non-conservative (Ser → Gly) substitution at 568. On the other hand, comparison of the deduced N-terminal sequence of the open reading frame of the cDNA insert of pAVP-3 and the N-terminus of the substrate-binding subunit of the V-PPase from *Vigna* (Maeshima and Yoshida, 1989) reveals 19 identities and 5 conservative substitutions within a span of 30 amino acid residues starting at position 3. These findings not only demonstrate that the open reading frame of the cDNA insert of pAVP-3 (and pAVP-1 and p-AVP-2) encodes the substrate-binding subunit of the V-PPase but also show, in accord with the immunological data (*Section 2,4*), that this subunit exhibits a high degree of sequence conservation between *Arabidopsis*, *Beta* and *Vigna*.

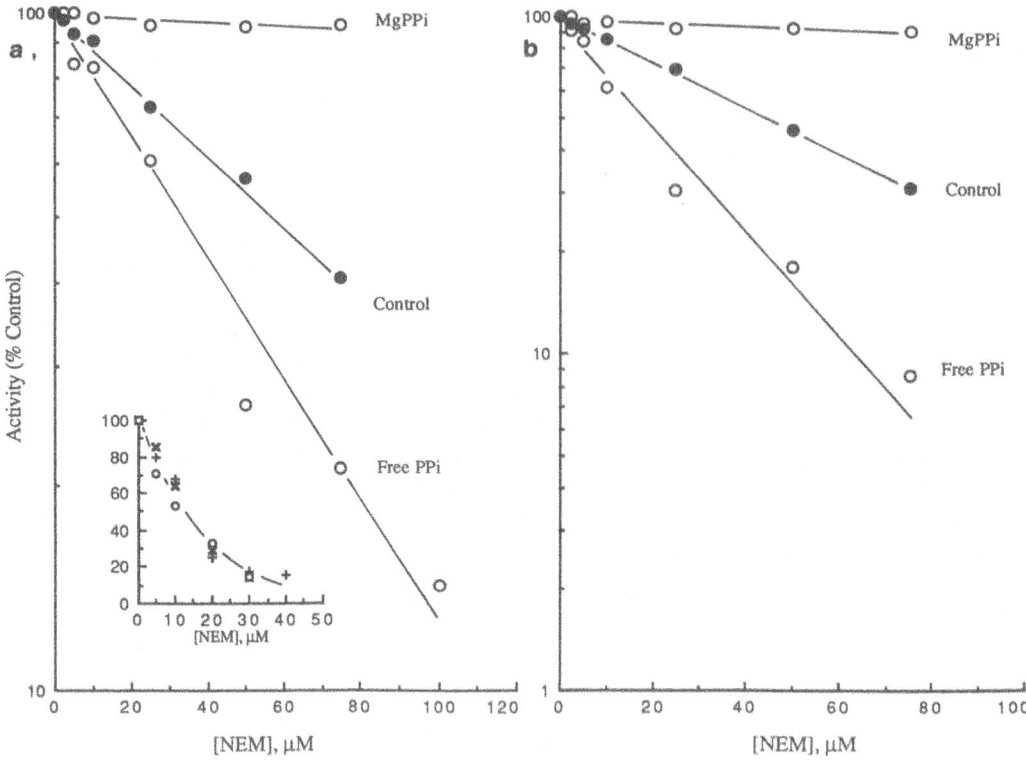

Figure 1A Kinetics of inhibition of vacuolar H^+-PPase from *Vigna* and *Beta* by NEM.
Main figure: Effects of no ligands ("Control"), free PP_i (0.3 mM Tris-PP_i) and $MgPP_i$ (0.3 mM Tris-PP_i + 1.3 mM $MgSO_4$) on inhibition of V-PPase of *Beta (Panel a)* and *Vigna (Panel b)* tonoplast vesicles by NEM.
Inset: Effect of membrane protein concentration on kinetics of inhibition of V-PPase of *Beta* tonoplast vesicles by NEM at membrane protein concentrations of 1 mg/ml (O), 2 mg/ml (x), and 4 mg/ml (+), respectively.

3.2   One gene, two membranes?

Genomic Southern analysis suggest that the gene encoding the 81 kDa polypeptide is present in a single copy in the genome of *Arabidopsis* (Sarafian *et al.*, 1991). On the other hand, subcellular fractionation of plant cells, suggests that the V-PPase, like the V-ATPase, co-resident on the same membrane, is present in at least two organellar membrane systems: those bounding the vacuole (Rea and Poole, 1985, Wang *et al.*, 1986) and those constituting the Golgi complex (Chanson, Fichmann, Spear, and Taiz, 1985). The results deriving from the molecular investigations consequently complement and extend the results of kinetic studies to indicate that the same polypeptide mediates PP_i-dependent H^+-translocation in both systems. Since the gene encoding the tonoplast H^+-PPase is present in a single copy in the genome of *Arabidopsis*, and it is likely that the labelled probe used for the genomic hybridisations would detect a second highly conserved H^+-PPase gene in the same organism, it would appear that the gene detected encodes the H^+-PPase of both vacuolar and Golgi membranes.

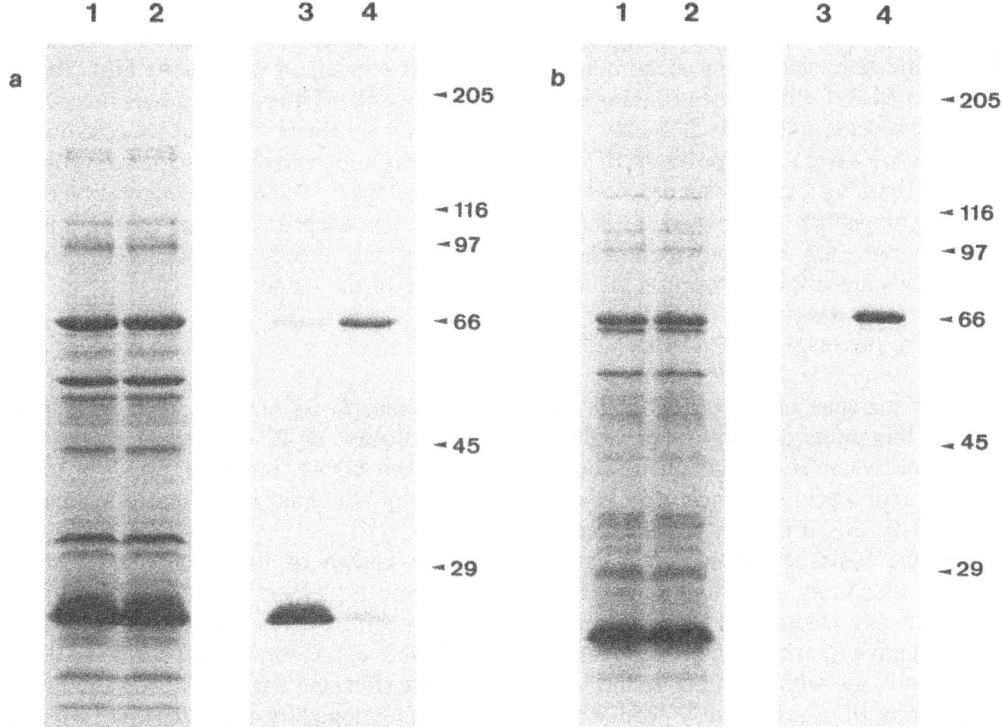

Figure 1B SDS-PAGE and fluorography of [^{14}C]-NEM-treated tonoplast vesicles from *Beta* and *Vigna*.
a. Coomassie blue-stained gel. *Lanes 1 and 2*, *Beta* tonoplast vesicles (50 μg) labelled with 50 μM [^{14}C]-NEM in the presence of MgPP$_i$ (*lane 1*) or free PP$_i$ (*lane 2*). *Lanes 3 and 4*, *Vigna* tonoplast vesicles labelled with [^{14}C]-NEM in the presence of MgPP$_i$ (*lane 3*) or free PP$_i$ (*lane 4*).
b. Fluorogram of [^{14}C]-NEM-treated tonoplast vesicles. *Lanes 1 and 2*, *Beta* tonoplast vesicles labelled with [^{14}C]-NEM in the presence of MgPP$_i$ (*lane 1*) or free PP$_i$ (*lane 2*). *Lanes 3 and 4*, *Vigna* tonoplast vesicles labelled with [^{14}C]-NEM in the presence of MgPP$_i$ (*lane 3*) or free PP$_i$ (*lane 4*).
All samples were pre-treated with 50 μM [^{14}C]-NEM + MgPP$_i$ (= 1.3 mM MgSO$_4$ + 0.3 mM Tris-PP$_i$) before incubation with 50 μM [^{14}C]-NEM + free PP$_i$ (= 0.3 mM Tris-PP$_i$) or MgPP$_i$ (= 1.3 mM MgSO$_4$ + 0.3 mM Tris-PP$_i$).

3.3 Topography of V-PPase

Computer-assisted hydropathy plots of the *Arabidopsis* V-PPase amino acid sequence establish that the cDNA insert of pAVP-3 encodes an extremely hydrophobic, integral membrane protein (Figure 2). Membrane associated ∝-helices were determined by hydrophobic moment analysis (Eisenberg, Schwarz, Komaromy, and Wall, 1984) and some of the secondary structural characteristics of the hydrophilic sequences was assigned by the Garnier method (Garnier, Osguthorpe, and Robson, 1978). The model contains 13 trans-membrane spans, all of which are amphipathic ("multimeric") and may interact to stabilise the structure through the formation of inter-helical H-bonds and salt-bridges. In addition, several of the

31

hydrophilic domains are characterised by the presence of clusters of charged residues which may participate in anion (PP_i^{4-}, $MgPP_i^{2-}$) or cation (Mg^{2+}, Ca^{2+}, K^+) binding. Thus, the hydrophilic segment linking membrane spans I and II contains 4 contiguous Glu residues (positions 64-67), the segment linking spans IV and V contains 4 acidic residues in a stretch of 8 amino acids (positions 222-229), the segment between spans VIII and IX contains the sequence ArgXArgXArg (positions 525-529) and the penultimate hydrophilic domain, linking spans XII and XIII, contains a preponderance of Lys residues. The overall orientation of the 81 kDa polypeptide is depicted, as shown in Figure 2, in accord with the "positive-inside rule" wherein, for most polytopic membrane proteins, the majority of positively charged amino acids are disposed towards the cytoplasmic face of the membrane.

3.4   Unique origins of V-PPase

Database searches of the sequence of the V-PPase confirm its novelty. No homologies between this pump and any other sequenced ion translocase or PP_i-dependent enzyme are discernible when either the nucleotide sequence of the cDNA insert of pAVP-3 or the deduced amino acid sequence of polypeptide encoded by the clone are searched against the GENBANK, GENPEPT or Swiss Protein Databases.

On the basis of the sequence data and what is known of its subunit composition, substrate-specificity and inhibitor-sensitivity, the V-type H^+-PPase must be ascribed to a completely new category of H^+-translocase. The type-specific inhibitors azide, orthovanadate and bafilomycin, which selectively and strongly inhibit F-, P- and V-type H^+-ATPases, respectively, are without effect on the V-type H^+-PPase (Rea and Sanders, 1987) and non of the subunits of F-, P- and V-type ATPases are immunologically cross-reactive with the $MgPP_i$-binding subunit of the enzyme (Rea, P.A., unpublished; Sarafian, V., and Poole, R.J., unpublished).

Close phylogenic links between the V-PPase and soluble PPases are also unlikely. All known soluble PPases have subunit sizes different from the tonoplast H^+-PPase - 20,000 Da for the enzymes from prokaryotes (Lahti, Pitkaranta, Valve, Ilta, Kukko-Kalske, and Heinonen, 1988) and 32,000-42,000 Da for the enzymes from eukaryotes (e.g. Kolakowski, Schloesser, and Cooperman, 1988) - and none of the known sequences for soluble PPases,[*Arabidopsis*, (Kieber and Signer, 1991), *E. coli*, (Lahti *et al.*, 1988), *Kluveromyces lactis*, (Stark and Milner, 1989), *Saccharomyces cerevisiae*, (Kolakowski *et al.*, 1988)] align with the deduced sequence of the V-PPase. Similarly, the recently cloned catalytic M_r 28,000-30,000 subunit of the membrane-bound mitochondrial ("F-") H^+-PPase (Lundin, Baltsheffsky, and Ronne, 1991) may be eliminated as a homologue. The subunit from *Saccharomyces* is 49% identical to the soluble PPase from the same source (Lundin, *et al.*, 1991) and shows no sequence identities with the V-PPase from *Arabidopsis*. Also, the corresponding enzyme from rat liver mitochondria does not cross-react with antibody raised against the V-type H^+-PPase from *Vigna* (Maeshima, 1991).

Evaluation of a potentially more promising evolutionary relationship between the tonoplast H^+-PPase and the reversible H^+-translocating PPase (H^+-PP_i synthase) of phototrophic bacteria will be contingent on the acquisition of sequence data from the latter category of enzyme. The existence of a H^+-PP_i synthase on the energy-coupling membranes of phototrophic bacteria, notably *Rhodosprillum rubrum*, has been known for some time, but only recently has it been shown that this translocase is an integral membrane protein with an apparent M_r of 56,000 (Nyren, Nore, and Strid, 1991). Two features of the M_r 56,000 polypeptide are significant: it is immunologically cross-reactive with the $MgPP_i$-binding sub-unit of the tonoplast H^+-PPase (Kim, E.J., Kim, Y., and Rea, P.A., unpublished) and, unlike the M_r 28,000-30,000 peripheral catalytic subunit of the mitochondrial H^+-PPase (Lundin *et al.*, 1991), the M_r 56,000 subunit, alone, is capable of mediating both $MgPP_i$ hydrolysis and

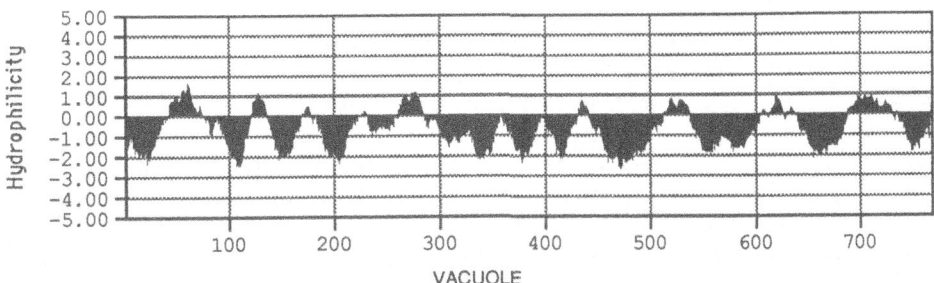

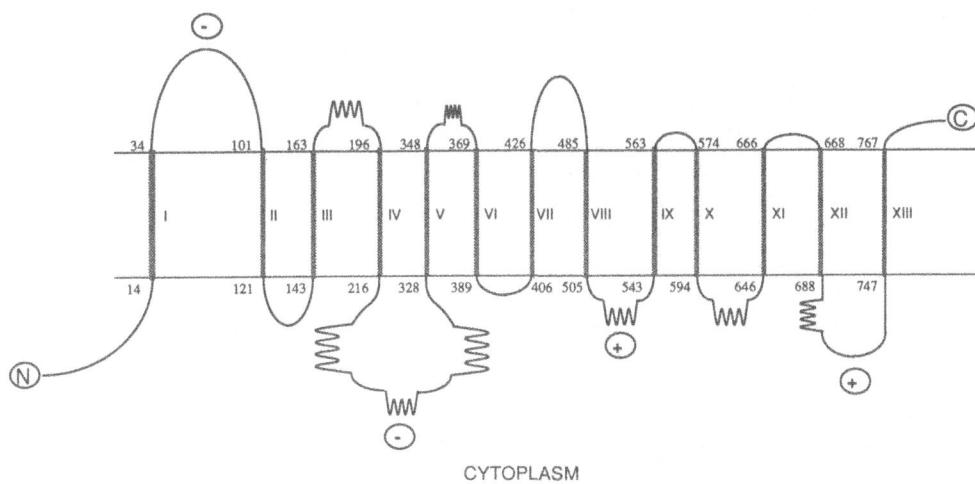

Figure 2 Computer-assisted hydrophilicity plot and topographic model of *Arabidopsis* tonoplast V-PPase amino acid sequence. The plots were calculated according to the Kyte-Doolittle method of MacVector over a running window of 20 amino acid residues. Values above the median represent hydrophilic segments; values below the median represent hydrophobic segments. The numbers I through XIII represent the putative membrane-spanning segments depicted in the tentative topographic model of the V-PPase. The transmembrane segments - all of which appear to be multimeric on the basis of their hydrophobic moment - were predicted by the HELIXMEM program of PC/GENE. The structure of the non-transmembrane regions was examined according to the secondary structure predictions of Garnier *et al.*, (1978) using both the GARNIER and GGBSM programs of PC/GENE. The structures indicated are: , ∝-helix; —, random coil; ⊖, clusters of negative charge; ⊕ clusters of positive charge; N, amino-terminus; C, carboxy-terminus.

H$^+$-translocation (Nyren *et al.*, 1991). Therefore, exciting is the possibility that the H$^+$-PPases from *Rhodosprillum* and plant vacuoles are structurally and functionally homologous. Genomic screens of *Rhodosprillum* using the cDNA insert of pAVP-3, and derivatives thereof, as probes are currently in progress in our laboratory to examine the validity of this proposal.

3.5 Putative catalytic site(s)

Lack of sequence homology does not automatically preclude functional analogies. The potential importance of this point is evident when it is noted that direct sequence comparisons between the V-PPase and other known PPase are complicated by the fact that clones of the soluble PPase from a range of organisms including *Arabidopsis, E. coli, Kluyveromyces, Saccharomyces* and *Schizosaccharomyces* (Lahti, Kolakowski, Heinonen, Vihinen, Pohjanoksa, and Cooperman, 1990) demonstrate only modest overall sequence identities of 20-27%, although functionally important residues show pronounced conservation. Crystallographic studies of the soluble PPase from *Saccharomyces* have identified 17 residues thought to be involved in Mg^{2+} and PP$_i$ binding and 11-16 of these residues (depending on alignment procedure) are conserved between the enzymes from *Arabidopsis, E. coli, Kluyveromyces, Saccharomyces* and *Schizosaccharomyces* (Lahti *et al.*, 1990). Eight of the active site residues of soluble PPases fall into two configurations - E̲XXXXXXXK̲X̲E̲ and D̲EXEXD̲XK̲XXXXD - at positions 48 and 146, respectively, in the sequence of the enzyme from *Saccharomyces* (Lahti *et al.*, 1990). Therefore, it is significant that variants of these two motifs - D̲XXXXXXXK̲XE and D̲XXXXD̲XK̲XXXXD - are present at positions 257 and 119, respectively, in the deduced amino acid sequence of the *Arabidopsis* V-PPase (Figure 3). If the gross orientation of the tonoplast H$^+$-PPase is as depicted in Figure 2, the motif DXXXXDXKXXXXD lies within the second hydrophilic domain connecting transmembrane spans II and III whereas the motif DXXXXXXXKXE lies within the fourth extramembraneous loop connecting spans IV and V. Hence, both motifs would be expected to be located on the cytoplasmic face of the membrane, the minimum condition to be fulfilled for direct interaction with cytosolic ligands. Moreover, the putative MgPP$_i$-binding motif at position 119 contains one Cys residue and is immediately flanked by another. Therefore, it is tempting to speculate that the pronounced sensitivity of the V-PPase to MgPP$_i$-protectable, free PP$_i$-potentiated inhibition and covalent modification by the sulfhydryl reagent N-ethylmaleimide (Britten *et al.*, 1989; Rea *et al.*, 1992) is attributable to alkylation of one or both of these Cys residues. Thus, while the V-PPase and soluble PPases appear to be remote evolutionarily, they may share convergent sequence motifs associated with the need for both classes of enzyme to interact with the same substrates and co-factors.

3.6 How many V-PPase polypeptides?

The MgPP$_i$-binding subunit of the V-PPase appears to be the sole polypeptide constituting the functional enzyme complex. This component, alone, co-purifies with PPase activity during detergent-solubilisation and chromatography (Britten *et al.*, 1989; Maeshima and Yoshida, 1989; Rea *et al.*, 1992) and is the only polypeptide of the tonoplast vesicles susceptible to MgPP$_i$-protectable, free PP$_i$-potentiated labelling by [^{14}C]-NEM (Britten *et al.*, 1989; Rea *et al.*, 1992). Accordingly, selective purification of the MgPP$_i$-binding subunit of the tonoplast H$^+$-PPase from *Vigna* and its incorporation into artificial liposomes results in the reconstitution of both MgPP$_i$ hydrolysis and MgPP$_i$-dependent H$^+$-translocation (Britten, C.J., and Rea, P.A., in preparation). Together with the amino acid sequence data derived from pAVP-3, which show that the MgPP$_i$-binding subunit of the tonoplast H$^+$-PPase is highly hydrophobic and possesses multiple trans-membrane spans, these findings strongly suggest that subunits in addition to the 81 kDa polypeptide encoded by AVP-3 need not be invoked to account for the capacity of the V-PPase for PP$_i$-energised trans-tonoplast H$^+$-translocation. This conclusion is corroborated by recent estimates of the target size ("functional mass") of the tonoplast H$^+$-PPase during (uncoupled) substrate hydrolysis which yield values of between 91 and 96 kDa (Pugliarello *et al.*, 1991; Sarafian, V., Potier, M., and Poole, R.J., in preparation); values correspondent with a deduced subunit mass of 81 kDa for the protein encoded by AVP-3. Although the PPase elutes as a dimer during gel filtration chromatography (Rea and Poole, 1986; Britten *et al.*, 1989; Sarafian and Poole, 1989) and may function as a homotrimer during H$^+$-translocation (Sarafian, V., Potier, M., and Poole, R.J., in preparation), the data available do not contradict the possibility that the V-PPase has been cloned in its entirety.

		*	* *	** * * *
PPA-1	Sce	48-**E**IPRWTNA**K**L**E**		146-**D**EGET**D**W**K**VIAI**D**
	Kla	**E**IPRWTNA**K**L**E**		**D**EGET**D**W**K**VIAI**D**
	Spo	**E**IPRWTQA**K**L**E**		**D**EGET**D**W**K**VIVI**D**
	Ara	**E**APTVFNC**K**L**E**		**D**QGEK**D**D**K**IIAVC
PPA-2	Sce	**E**VPRWTTG**K**F**E**		**D**DGEL**D**W**K**VIVI**D**
AVP	Ara	257-**D**VGADLVG**K**I**E**		119-**E**GFST**D**N**K**PCTY**D**
Consensus		(**D/E**)XXXXXXX**K**X**E**		(**D/E**)XXXX**D**X**K**XXXX**D**

Figure 3 Alignment of putative catalytic amino acid residues of soluble PPases, mitochondrial H$^+$-PPase and V-PPase (AVP). Putative active site residues (Lahti *et al.*, 1990) are indicated by asterisks. Residues of like-charge, common to all six sequences, are shown in boldface. The numbering of the residues is that of the mature *Saccharomyces* soluble PPase (PPA-1) protein. Shown are the sequences for the soluble PPases (PPA-1) from *Saccharomyces cerevisiae (Sce), Kluyveromyces lactis (Kla), Schizosaccharomyces pombe (Spo)* and *Arabidopsis (Ara)*, the mitochondrial H$^+$-PPase (PPA-2) from *S. cerevisiae* and the V-PPase (AVP) from *Arabidopsis*.

4.   Closing Remarks

To our knowledge, the findings reported here represent the first isolation and sequence analysis of any V-PPase clone. As such, they make this unique H$^+$ pump amenable to detailed genetic analysis for the first time. The full physiological significance of the V-PPase *in vivo* might therefore be addressed by gene transformation techniques. Two overriding questions are the involvement of the V-PPase in cellular PP$_i$ regulation and vacuolar K$^+$ transport. A remarkable characteristic of cellular PP$_i$ levels in plants is their invariancy. PP$_i$ levels do not change during light-dark transitions (Weiner, Stitt and Heldt, 1987), during rapid changes in respiration rate or when tissues are subjected to anoxia or respiratory poisons (Dancer and ap Rees, 1988). The operation of a highly effective PP$_i$-stat is therefore implicated. What are the roles of the V-PPase and the recently cloned soluble PPase of

Arabidopsis (Kieber and Signer, 1991) in cellular PP$_i$-stasis and how do the two enzymes interact? At the level of plant mineral nutrition, recent patch clamp studies on intact, isolated vacuoles strongly suggest that rather than merely acting as a supplementary H$^+$ pump, the H$^+$-PPase may also catalyse K$^+$-H$^+$ symport into the vacuole (Davies, Rea and Sanders, 1991). Therefore, might the V-PPase be instrumental for cell turgor regulation in general, and stomatal guard cell inflation, in particular? The availability of cDNAs encoding the V-PPase should now enable the construction of transgenic plants that either express antisense transcripts of the AVP gene or ectopically over-express the gene for investigations of the phenotypic consequences of perturbed V-PPase gene expression on cellular PP$_i$ and K$^+$ homeostasis.

The abundance and ubiquity of the tonoplast PPase of plant cells challenges conventional notions of the role of PP$_i$ in biological energy transduction. While speculations concerning the evolutionary origins of the V-PPase await sequence data from candidate homologues, the unique status of PP$_i$ as the limiting case of a high energy phosphate make the results deriving from such investigations all the more intriguing. The results of experiments designed to simulate primitive earth conditions suggest that the formation of PP$_i$ and other polyphosphates may have been pivotal for prebiotic evolution: these molecules have been shown to serve as catalysts for the phosphorylation of adenosine, AMP and ADP and the synthesis of peptides (Yamanaka, Inomata, and Yamagata, 1988). Moreover, since the synthesis of PP$_i$ from precipitated orthophosphate at low temperatures is practicable (Hermes-Lima, 1990) and these conditions also approximate to those for template-directed oligomerisation of nucleotides (Acevedo and Orgel, 1986), it is conceivable that the generation of PP$_i$ and the synthesis of bio-polymers were necessary concomitants for the emergence of life. Any system which could harness the energy contained in PP$_i$ would consequently be at a replicative advantage. Is the energy-conserving V-PPase and other enzymes of its type, therefore a vestige of life's prebiotic origins before the limits on the general utility of PP$_i$ in living systems were brought to bear by the emergence of the more specific and recognizable structure of ATP (and other nucleotides) for nucleic acid synthesis (gene replication and transcription), adenylation, pyrophosphorylation and signal transduction?

Acknowledgements

This work was supported by the National Science Foundation (Grant No. DCB-9005330), the Department of Energy (Grant No. DE-FG02-91ER20055), the University Research Foundation, University of Pennsylvania and a grant from the Biological Sciences Research Group, University of Pennsylvania awarded to Philip Rea. Vahe Sarafian was partly supported by the Natural Sciences and Engineering Research Council, Canada by a grant awarded to Ronald Poole.

REFERENCES

ACEVEDO, O.L., and ORGEL, L.E., 1986 Template-directed oligonucleotide ligation on hydroxylapatite. *Nature*, **321**, 790-792.

AEBERSOLD, R.H., LEAVITT, J., SAAVEDRA, R.A., HOOD, L.E., and KENT, S.B.H., 1987. Internal amino acid sequence analysis of proteins separated by one- and two-dimensional gel electrophoresis after *in situ* protease digestion on nitrocellulose. *Proceedings of the National Academy of Sciences, USA*. **84**, 6970-6974.

BENNETT, A.B., O-NEILL, S.D., and SPANSWICK, R.M., 1984. H$^+$-ATPase activity from storage tissue of *Beta vulgaris*. I. Identification and characterisation of an anion-sensitive H$^+$-ATPase. *Plant Physiology*, **74**, 538-544.

BRITTEN, C.J., TURNER, J.C., and REA, P.A., 1989. Identification and purification of substrate-binding subunit of higher plant H^+-translocating inorganic pyrophosphatase. *FEBS Letters*, **256**, 200-206.

CHANSON, A., FICHMANN, J., SPEAR, D., and TAIZ, L., 1985. Pyrophosphate-driven proton transport by microsomal membranes of corn coleoptiles. *Plant Physiology*, **79**, 159-164.

CHANSON, A., and PILET, P.E., 1989. Target molecular size and sodium dodecyl sulfate-polyacrylamide gel electrophoresis analysis of the ATP- and pyrophosphate-dependent proton pumps from maize root tonoplast. *Plant Physiology*, **90**, 934-938.

CHURCHILL, K.A., and SZE, H., 1983. Anion-sensitive H^+ pumping ATPase in membrane vesicles from oat roots. *Plant Physiology*, **71**, 610-617.

DANCER, J.E., and AP REES, T., 1989. Relationship between pyrophosphate : fructose-6-phosphate 1-phosphotransferase, sucrose breakdown and respiration. *Journal of Plant Physiology*, **135**, 197-206.

DAVIES, J.M., REA, P.A., and SANDERS, D., 1991. Vacuolar proton-pumping pyrophosphatase in *Beta vulgaris* shows vectorial activation by potassium . *FEBS Letters*, **278**, 66-68.

EISENBERG, D., SCHWARZ, E., KOMAROMY, M., and WALL, R., 1984. Analysis of membrane and surface protein sequences with the hydrophobic moment plot. *Journal of Molecular Biology*, **179**, 125-142.

FRAICHARD, A., MAGNIN, T., and PUGIN, A., 1991. Properties and purification of a proton translocating pyrophosphatase from the tonoplast of *Acer pseudoplatanus*. *Plant Physiology*, **96**, S1049.

GARNIER, J., OSGUTHORPE, D.J., and ROBSON, B., 1978. Analysis of the accuracy and implications of simple methods for predicting the secondary structure of globular proteins. *Journal of Molecular Biology*, **120**, 97-120.

HAROLD, F.M., 1986. *The Vital Force: A Study of Bioenergetics*. Freeman, New York.

HERMES-LIMA, M., 1990. Model for prebiotic pyrophosphate formation: condensation of precipitated orthophosphate at low temperature in the absence of condensing or phosphorylating agents. *Journal of Molecular Evolution*, **31**, 353-358.

KARLSSON, J., 1975. Membrane-bound potassium and magnesium ion stimulated inorganic pyrophosphatase from roots and cotyledons of sugar-beet (*Beta vulgaris* L.). *Biochimica et Biophysica Acta*, **399**, 256-363.

KIEBER, J.J., and SIGNER, E.R., 1991. Cloning and characterisation of an inorganic pyrophosphatase gene from *Arabidopsis thaliana*. *Plant Molecular Biology*, **16**, 345-348.

KOLAKOWSKI, L.F., SCHLOESSER, M., and COOPERMAN, B.S., 1988. Cloning, molecular characterisation and chromosome localisation of the inorganic pyrophosphatase (PPA) gene from *S. cerevisiae*. *Nucleic Acids Research*, **22**, 10441-10452.

KRISHNAN, V.A., and GNANAM, A., 1988. Properties and regulation of Mg^{2+}-dependent chloroplast inorganic pyrophosphatase from *Sorghum vulgare* leaves. *Archives of Biochemistry and Biophysics*, **260**, 277-284.

KULAKOVSKAYA, T.V., LICHKO, L.P., and OKOROKOV, L.A, 1989. Membrane-bound pyrophosphatase is second vacuolar H^+ pump in yeast. *Soviet Plant Physiology*, **36**, 5-10.

LAHTI, R., PITKARANTA, T., VALVE, E., ILTA, I., KUKKO-KALSKE, E., and KEINONEN, J., 1988. Cloning and characterisation of the gene encoding inorganic pyrophosphatase of *Escherichia coli* K-12. *Journal of Bacteriology*, **170**, 5901-5907.

LAHTI, R., KOLAKOWSKI, F.L., HEINONEN, J., VIHINEN, M., POHJANOKSA, K., and COOPERMAN, B.S., 1990. Conservation of functional residues between yeast and *E. coli* inorganic pyrophosphatases. *Biochimica et Biophysica Acta*, **1038**, 338-345.

LUNDIN, M., BALTSHEFFSKY, H., and RONNE, H., 1991. Yeast *PPA2* gene encodes a mitochondrial inorganic pyrophosphatase that is essential for mitochondrial function. *Journal of Biological Chemistry*, **266**, 12168-12172.

MANOLSON, M.F., PERCY, J.M., APPS, D.K., XIE, X.-S., STONE, D.K., HARRISON, M., CLARK, D.J., and POOLE, R.J., 1989. Evolution of vacuolar H^+-ATPase: immunological relationships of the nucleotide-binding subunits. *Biochemical Cell Biology*, **67**, 306-310.

MAESHIMA, M., 1991. H^+-translocating inorganic pyrophosphatase of plant vacuoles. Inhibition by Ca^{2+}, stabilization by Mg^{2+} and immunological comparison with other inorganic pyrophosphatases. *European Journal of Biochemistry*, **196**, 11-17.

MAESHIMA, M., and YOSHIDA, S., 1989. Purification and properties of vacuolar membrane proton-translocating inorganic pyrophosphatase from mung bean. *Journal of Biological Chemistry*, **264**, 20068-20073.

MITCHELL, P., 1961. *Chemiosmotic Coupling in Oxidative and Photosynthetic Phosphorylation*. Glynn Research, Bodmin, Cornwall, UK.

NELSON, N., and TAIZ, L., 1989. The evolution of H$^+$-ATPases. *Trends in Biochemical Sciences*, **14**, 113-116.

NYREN, P., NORE, B., and STRID, A., 1991. Proton-pumping *N,N'*-dicyclohexylcarbodiimide sensitive inorganic pyrophosphate synthase from *Rhodospirillum rubrum*: purification, characterisation and reconstitution. *Biochemistry*, **30**, 2883-2887.

REA, P.A., BRITTEN, C.J., and SARAFIAN, V., 1992. Common identity of substrate-binding subunit of vacuolar H$^+$-translocating inorganic pyrophosphatase of higher plant cells. *Journal of Biological Chemistry*, in press.

REA, P.A., and POOLE, R.J., 1985. Proton-translocating inorganic pyrophosphatase in red beet (*Beta vulgaris* L.) tonoplast vesicles. *Plant Physiology*, **77**, 46-52.

REA, P.A., and POOLE, R.J., 1986. Chromatographic resolution of H$^+$-translocating pyrophosphatase from H$^+$-translocating ATPase of higher plant tonoplast. *Plant Physiology*, **81**, 126-129.

REA, P.A., and SANDERS, D., 1987. Tonoplast energisation: two pumps, one membrane. *Physiologia Plantarum*, **71**, 131-141.

SARAFIAN, V., and POOLE, R.J., 1989. Purification of an H$^+$-translocating inorganic pyrophosphatase from vacuolar membranes of red beet. *Plant Physiology*, **91**, 34-38.

SARAFIAN, V., KIM, Y., POOLE, R.J., and REA, P.A., 1992. Molecular cloning and sequence of cDNA encoding the pyrophosphate-energised vacuolar membrane proton pump (H$^+$-PPase) of *Arabidopsis thaliana*. *Proceedings of the National Academy of Sciences, USA*. in press.

STARK, M.J.R., and MILNER, J.S., 1989. Cloning and analysis of the *Kluyveromyces lactis TRP1* gene: a chromosomal locus flanked by genes encoding inorganic pyrophosphatase and histone H3. *Yeast*, **5**, 35-50.

WALKER, R.R., and LEIGH, R.A., 1981. Mg^{2+}-dependent, cation-simulated inorganic pyrophosphatase associated with vacuoles isolated from storage roots of red beet (*Beta vulgaris* L.). *Planta*, **153**, 150-155.

WANG, Y., LEIGH, R.A., KAESTNER, K.H., and SZE, H., 1986. Electrogenic H$^+$-pumping pyrophosphatase in tonoplast vesicles of oat roots. *Plant Physiology*, **81**, 497-502.

WEINER, H., STITT, M., and HELDT, H.W., 1987. Subcellular compartmentation of pyrophosphate and alkaline pyrophosphatase in leaves. *Biochimica et Biophysica Acta*, **893**, 13-21.

WHITE, P.J., MARSHALL, J., and SMITH, J.A.C., 1990. Substrate kinetics of the tonoplast H$^+$-translocating inorganic pyrophosphatase and its activation by free Mg^{2+}. *Plant Physiology*, **93**, 1063-1070.

YAMANAKA, J., INOMATA, K., and YAMAGATA, Y., 1988. Condensation of oligoglycines with trimeta- and tetrametaphosphate in aqueous solutions. *Origins of Life*, **18**, 165-178.

Studies on the Higher Plant Calmodulin-Stimulated ATPase

**David E. Evans[1,2], Per Askerlund[2], Joy M. Boyce[2],
Sally-Anne Briars[2,3], David Coates[4], Janice Coates[4],
David T.Cooke[5] and Frederica L. Theodoulou[2]**

1. Author for correspondence

2. Department of Plant Sciences
    University of Oxford
    South Parks Road
    Oxford, OX1 3RA, UK

3. Present address: Unilever Research
    Colworth House
    Sharnbrook
    Bedford, MK44 1LQ, UK

4. Department of Pure and Applied Biology
    University of Leeds
    Leeds, LS2 9JT, UK

5. Department of Plant Sciences
    IACR, Long Ashton Research Station
    Long Ashton
    Bristol, BS18 9AF, UK

1.   Introduction

Plant cells regulate their cytoplasmic calcium concentration ($[Ca]_{cyt}$) to lower than 1.0 μM, despite a steep gradient (probably around 1000 x ; Meir, Juniper and Evans, 1991) favouring Ca^{2+} influx into the cell (reviewed by Evans, Briars and Williams, 1991).   This low concentration is maintained by the action of active calcium transport systems, located at the plasma membrane (e.g., Butcher and Evans, 1987a,b; Rasi-Caldogno, Pugliarello, DeMichelis, 1987; Gräf and Weiler, 1989), endoplasmic reticulum (e.g., Bush and Sze, 1986; Brauer, Schubert and Tsu, 1990; Bush, Biswas and Jones, 1989) and tonoplast (e.g., Schumaker and Sze, 1985, Blumwald and Poole, 1986; and see also Evans, 1988 and Evans

Transport and Receptor Proteins of Plant Membranes
Edited by D.T. Cooke and D.T. Clarkson, Plenum Press, New York, 1992

39

et al., 1991, Briskin, 1991 for reviews). Plant active calcium transporters fall into two categories; P-type ion translocating ATPases and calcium-proton antiporters; it has been suggested that the former are located at the plasma membrane and endoplasmic reticulum, while the latter is located at the tonoplast, although alternative locations have also been suggested (reviewed by Evans *et al.*, 1991).

Early reports of calcium transport by plant cell membranes indicated the presence of a calmodulin-stimulated calcium transporter in a mixed (microsomal) membrane fraction from *Zea mays* L. (Dieter and Marmé, 1980, 1981a,b, 1983). This activity showed properties similar to the primary active calcium transporter of the plasma membrane of many mammalian cell types (Table 1); it showed a nearly identical dose-response curve for calmodulin, similar pH optimum, used MgATP as substrate and (in membrane preparations) showed activity appropriate for a pump acting to expel calcium from the cytosol.

Table 1 Comparison of properties of higher plant and mammalian calmodulin (CaM)-stimulated ATPase.

	Plant	Erythrocyte
M_r	c 140,000	138,000 $\pm$ 4,000
CaM stimulation?	Yes; 50%-0.1-0.2μM	Yes; 50%-c. 0.1μM
Mechanism	P-type ATPase	P-type ATPase
Common antigenicity	Yes	Yes
Substrate	MgATP	MgATP
Activity range	pCa 5-6	pCa 5-6
pH optimum	7.2	7.2
Membrane location	?	plasma membrane
Action	Pumps Ca^{2+} from cytosol	Pumps Ca^{2+} from cytosol

Calmodulin-stimulated ATPase activity could also be purified by calmodulin-affinity chromatography (Dieter and Marmé, 1981a) a technique applied routinely to the purification of the mammalian calmodulin-stimulated calcium pump (Niggli, Penniston and Carafoli, 1979). Our own work has built upon these early findings of Dieter and Marmé. Using refinements of the purification scheme they employed, a partially-purified active calmodulin-stimulated ATPase was obtained using microsomal membranes from dark-grown *Zea mays* L. seedlings (Briars, Kessler and Evans, 1988; Briars, Dewey and Evans, 1989). Gel electrophoresis of this fraction revealed that a major polypeptide of 140,000 M_r was present, together with a number of other, smaller polypeptides. This 140,000 M_r polypeptide was recognised specifically by an affinity-purified polyclonal antibody raised against the erythrocyte calmodulin-stimulated Ca^{2+} pump (Briars, Kessler and Evans, 1988), strongly suggesting that it is the higher plant calmodulin-stimulated (CaM) ATPase. This was further suggested by the similar M_r of the erythrocyte pump (135,000; Niggli *et al.*, 1979).

P-type ion-translocating ATPases form a phosphorylated intermediate state as part of their reaction mechanism. Such phosphorylated intermediates are detectable as radioactive bands on acidic polyacrylamide gels if membrane or purified enzyme preparations are incubated with high specific activity γ-[^{32}-P]-ATP before separation. Studies (Briars and Evans, 1989) on plant microsomes and on the purified calmodulin-stimulated ATPase suggest that the 140,000 M_r polypeptide does form such a phosphorylated intermediate, having properties appropriate for a P-type calcium pump (inhibition of dephosphorylation by lanthanum; sensitivity to hydroxylamine and chasing with unlabelled ATP), providing further confirmation that *Zea mays* microsomes contain a P-type, calmodulin-stimulated Ca^{2+} pump similar in size and properties to that of the mammalian plasma membrane Ca^{2+} pump.

In spite of the above apparent homology with a mammalian plasma membrane Ca^{2+} pump, some doubt remains as to the subcellular localisation of the calmodulin-stimulated ATPase. Table 2 summarises reports to date suggesting a variety of membrane types, purified by a variety of techniques showing direct and indirect calmodulin-stimulated ATP hydrolysis or calcium transport. However, a number of other reports (e.g., Butcher and Evans, 1987, Gräf and Weiler, 1989) have failed to show calmodulin-stimulated calcium transport in plasma membrane, even after reconstitution (Gräf and Weiler, 1990). Whilst failure to demonstrate activity does not necessarily indicate absence of activity (e.g., due to saturation of membranes with endogenous calmodulin; partial proteolysis of membrane pumps, etc.), it is clear that the subcellular location(s) of the calmodulin-stimulated calcium pump in higher plant cells remains an open question.

Functional reconstitution of purified or partially purified higher plant plasma membrane calcium pumps has been achieved by two laboratories (Gräf and Weiler, 1990; Kasai and Muto, 1991). calmodulin-stimulated was not demonstrated in either of these studies. In this paper, we describe the first functional reconstitution of the calmodulin-stimulated ATPase from higher plants together with preliminary experiments to investigate its membrane environment.

2.   Material and Methods

All chemicals and biochemicals were purchased from Sigma (Poole, Dorset, UK) or BDH (Poole, Dorset, UK) and were of the highest available grade unless specified otherwise. Dark-grown, 4/5 d maize (*Zea mays* L. cv Golden Bantam or LG 20-80) seedlings were grown as described previously (Briars, Kessler and Evans, 1988). Cauliflower inflorescences were carefully selected and purchased locally. Protein content was determined by a modified Bradford assay (Stoscheck, 1990).

3.   Purification of Maize Plasma Membrane

Plasma membranes were prepared from a microsomal fraction (10,000 - 50,000 x g pellet) by partitioning in a aqueous polymer two-phase system (Larsson, Widell and Kjellbom, 1987). The coleoptiles (90 g) were homogenised in a Braun homogeniser equipped with razor blades in place of the ordinary knives (Kannangara, Gough, Hansen, Rasmussen and Simpson, 1977).

The homogenisation medium (230 ml) contained 50 mM MOPS-KOH (pH 7.5), 330 mM sucrose, 5 mM Na$_2$EDTA, 0.2% (w/v) casein (boiled acid hydrolysate), 0.2% (w/v) BSA (Sigma, protease free), 0.6% (w/v) insoluble PVP, 5 mM DTT and 0.5 mM PMSF. The microsomal pellets were resuspended in 330 mM sucrose, 5 mM K-phosphate (pH 7.8), 1 mM DTT and 0.5 mM PMSF, homogenised in a glass/teflon homogeniser and added to a phase system giving a 36 g system with a final composition of 6.5% (w/w) Dextran T500, 6.5% polyethylene glycol 3350, 330 mM sucrose, 3 mM KC1, 5 mM K-phosphate (pH 7.8). Phase partitioning was carried out as described in Larsson *et al.* (1987).

The resulting plasma membranes (denoted U_3 and U_3') and intracellular membranes (L_2) as well as microsomes (Mic) were diluted several-fold with 25 mM TRIS-MES (pH 6.5), 330 mM sucrose, 0.1 mM Na_2EDTA, 1 mM DTT and pelleted at 100,000 x *g* for 1h. The final pellets were resuspended in the same buffer and stored at -70° C until analysed. Marker enzyme activities were assayed as follows:

K^+, Mg^{2+}-ATPase was determined as described by Widell and Larsson (1990).

Table 2   Summary of subcellular locations suggested for higher plant calmodulin-stimulated ATPases and calcium transport.

Authors	Tissue	Location	Activity
Dieter and Marmé (1980, 1981)	*Zea* coleoptiles	Microsome	ATPase and Ca^{2+} transport
Zocchi (1988)	*Zea* roots	Tonoplast PM	Ca^{2+} transport Ca^{2+} transport
Andreev *et al.*, (1989)	Sugar-beet taproot	Tonoplast	Ca/H antiport
Brauer *et al.*, (1990)	*Zea* root	ER	Ca^{2+} transport
Robinson *et al.*, (1988)	*Zea* leaves	PM	ATPase
Malatialy *et al.*, (1988)	Spinach leaves	PM	Ca^{2+} transport
Williams *et al.*, (1990)	Red Beet	PM	Ca^{2+} transport

Glucan synthase II was assayed by the method of Fredrikson and Larsson (1989), antimycin-A insensitive NADH-cytochrome *c* reductase as described in Askerlund, Larsson, and Widell, (1988) except that 1 mM NaN_3 was substituted for KCN.

Cytochrome *c* oxidase was assayed at 550 nm using a Perkin-Elmer 551S spectrophotometer. The assay mixture contained 40-60 µg protein, 40 µM cytochrome *c* (Sigma C7752), 25 mM Hepes-KOH, pH 7.2, 0.33 M sucrose and 0.025% (w/v) Triton X-100. Prior to measurement, cytochrome *c* was reduced by the addition of a small amount of sodium dithionite; the excess was removed by extensive aeration. The reaction was started by the addition of membrane.

4.   Assay of Calmodulin-Simulated ATPase and Calcium Transport

Ca^{2+}-uptake was measured at 35° C in a medium containing 25 mM KC1, 0.1% w/v BSA (Sigma A7030), 2.5 mM $MgCl_2$, 2.5 mM ATP, 50 μM $CaCl_2$ (1-2 Bq $^{45}CaCl_2$ pmol^{-1}) and c 1.5 μg reconstituted ATPase in 0.1 ml. Calmodulin (Sigma P2277) was added as indicated (Figure 4). Free calcium was estimated using an Orion calcium electrode and varied from 30-40 μM between experiments. The reconstituted ATPase was incubated with assay medium for 20-30 minutes prior to starting the assay with ATP. The reaction was stopped by adding 0.6 ml 25 mM MOPS-BTP (pH 7.2), 1 mM EGTA, 0.33 M sucrose and aliquots were filtered immediately through 0.20 μm filters. After washing the filters 4 times with stop solution, they were air dried and radioactivity estimated by scintillation counting. Hydrolytic activity of the reconstituted ATPase was measured under identical conditions to the Ca^{2+} uptake, in the absence or presence of 5 μM A23187 as indicated and the Pi released was measured using a modified Baginski procedure (Widell and Larsson, 1990).

5.   Triton X-114 Partition of Maize Plasma Membrane Proteins and Immuno-Localisation of Calmodulin

Triton X-114 partition of plasma membrane proteins was done by the methods described in Kjellbom, Larsson, Rochester and Andersson (1989). Calmodulin was detected in immunoblots using an anti-calmodulin antibody which was the gift of Mark Collinge (University of Edinburgh).

6.   Functional Reconstitution of Maize Calmodulin-Stimulated ATPase

Briefly, the enzyme was purified from microsomal membranes prepared as described for a microsomal pellet prior to two-phase partition (above), with the addition of 1 mM benzamidine-HCl to the homogenisation mixture. The pellet was resuspended in medium A (25 mM MOPS-BTP, pH 7.2, 0.33 M sucrose) with 0.5 M NaCl, 1 mM Na_2EDTA, 5 mM DTT, 0.5 mM PMSF and pelleted at 100,000 x g for 45 minutes. The washed microsomal pellet was resuspended in 3 ml medium A plus 5 mM DTT, 0.5 mM PMSF. The washed microsomes were solubilised in 22.5 ml medium A plus 0.5 M $CaCl_2$, 5 mM $CaCl_2$, 0.1% (w/v) Sigma soybean phosphatidylcholine type IV-S, 2 mM $MgCl_2$, 2 mM ATP, 40 μM leupeptin, 0.5 mM PMSF, 2 mM DTT, 250 mg Triton X-100 (Surfact-Amps, Pierce and Warriner, Chester, UK) with stirring for ten minutes at 4° C. Unsolubilised material was removed by centrifugation at 100,000 x g for 45 minutes. Affinity chromatography was done as described previously (Niggli, et $al.$, 1979, Briars, et $al.$, 1988) with the addition of 0.05% phospholipid. After 150 ml wash, the detergent was changed to 1 mM CHAPS for 50 ml, and the proteins eluted using 10 mM EGTA. Peak fractions were pooled, supplemented with DTT and $CaCl_2$ (to 2.5 mM). Reconstitution was done by detergent dialysis. A mixed Sigma soybean phosphatidylcholine preparation (type IV-S) was further purified according to Helmke and Howard (1987) to remove contaminating protein and oxidants and stored in choroform-methanol (2:1). Solvent was removed by rotary evaporation followed by lyophilisation and the lipid resuspended in medium B (medium A, but pH 7.2) to 26 mg/ml. CHAPS (40 mM) was added and the suspension sonicated to clarity. To initiate reconstitution, 1 part lipid was added to 3 parts column eluent (ca. 150 μg protein in 7.5 ml). The mixture was dialysed for 60 hours against 3 x 1 litre medium B plus 1 mM DTT, 0.5 mM PMSF. During the last buffer change, 20 g washed Amberlite XAD2 was added. The proteoliposomes were stored under nitrogen with 10 mM DTT added until use.

7.   Lipid Analysis

Analysis of sterols (cholesterol, campesterol, stigmasterol, sitosterol) and phospholipids (PS, PC, PE, PI) were done as previously described (Cooke, et al., 1989). Prior to lipid analysis, the calmodulin-stimulated ATPase was purified as described in Briars, Kessler, and Evans, 1988.

Results

8.   Isolation and Purification of Plasma Membranes from Maize

The plasma membrane obtained using phase partitioning ($U_3 + U_3'$) represented 10% of the total protein in the microsomal fraction and was highly enriched in the plasma membrane marker activities K^+, Mg^{2+}-ATPase (6 times enrichment) and glucan synthase II (ca. 4 times enrichment) compared to the microsomal fraction. This ATPase was sensitive to vanadate but insensitive to nitrate. The plasma membrane fraction was almost completely depleted in cytochrome c oxidase activity, a marker for the inner mitochondrial membrane. In contrast, the specific antimycin A-insensitive NADH-cytochrome c reductase activity of the plasma membrane fraction was 75% of that in the microsomal fraction. This is in agreement with earlier investigations (Larsson, Widell, Kjellbom, 1987; Robinson, Larsson, and Buckhout, 1988; Askerlund, et al., 1988) and is probably due to the presence of NADH-cyt b_5 reductase as well as cyt b_5 in plant plasma membranes (Askerlund, Laurent, Nakagawa, and Kader, 1991). By far the majority of the total antimycin A-insensitive NADH-cyt c reductase activity partitioned in the intracellular membrane fraction (L_2).

9.   Calmodulin-Stimulated ATPase Activity and Ca^{2+} Transport in Maize Plasma Membrane

Demonstration of calcium transport by two-phase partition purified plasma membrane requires vesicles orientated cytoplasmic face outwards. These can be obtained by freezing and thawing the purified vesicles repeatedly. Plasma membrane vesicles treated in this way showed active calcium transport but, after washing with high salt and EGTA, it was not possible to demonstrate calmodulin-stimulation (data not shown), even in conditions known to permit the assay of calmodulin-stimulation in microsomal membranes. Similarly, for ATP-hydrolysis, with vesicles permeabilised using Triton X-100, ATP-hydrolysis (0.05 μmol/min/mg protein) was not stimulated by the addition of calmodulin; if anything, a very slight inhibition was observed (data not shown).

10.   Triton X-114 Partition of Calmodulin in Maize PM Purified by Aqueous Two Phase Partition

One possible explanation for a lack of stimulation of ATP-hydrolysis or calcium transport observed in the aqueous two-phase partition purified plasma membrane was that such membranes contain high levels of endogenous calmodulin (Collinge and Trewavas, 1989). This is partially due to the amphipathic nature of calmodulin, which partitions into the membrane and partially due to the fact that the purification method relies on the preparation of right-side out vesicles, which may entrap material from the homogenate. Immuno-blotting indicated that large amounts of endogenous calmodulin were present in the purified plasma membrane (Figure 1).

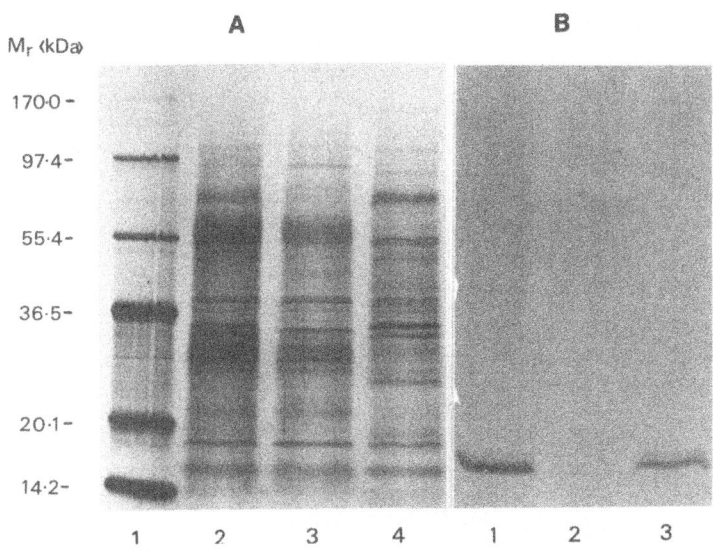

Figure 1 Effect of Triton X-114 partition on endogenous calmodulin in purified maize plasma membrane. Maize plasma membrane purified by aqueous two-phase partition (see 'Materials and Methods') was subjected to Triton X-114 partitioning as described in Kjellbom *et al.*, 1989. Figure 1a is a silver-stained SDS-polyacrylamide gel showing M_r markers (lane 1), purified plasma membrane before (lane 2) and after (lanes 3 and 4) Triton X-114 partition. Integral proteins partition into the Triton phase (lane 3) while peripheral proteins partition into the aqueous phase (lane 4). Due to its failure to silver stain, calmodulin is not visible in lanes 2-4; however, the "Western" blot is shown in Figure 1b done using an anti-calmodulin antibody (see 'Materials and Methods') illustrates that calmodulin is present in untreated plasma membrane (lane 1) and in the peripheral protein fraction (lane 3) but strongly depleted in the integral protein fraction (lane 2).

After Triton X-114 partition (Kjellbom, *et al.*, 1989), calmodulin was heavily depleted from the intrinsic membrane protein phase (TX-114 phase), but present in the peripheral protein phase (water phase; Figure 1). This calmodulin behaved as a peripheral membrane protein. After Triton X-114 partition, the fraction containing intrinsic membrane proteins (and now depleted in calmodulin) showed calmodulin-stimulated ATPase activity (Figure 2). The extent of stimulation was about 9%; however, it should be borne in mind that the basal activity would include other ATP hydrolysing enzymes.

11. Reconstitution of the Affinity-Purified Calmodulin-Stimulated ATPase from Cauliflower

Reconstitution strategy: Because it demonstrated greater calmodulin-stimulated ATPase activity, a microsomal fraction was used as starting material for purification in reconstitution experiments. Microsomes from dark gown maize and from cauliflower florets were found to be good sources of membrane for use in subsequent purification. The data presented here used cauliflower as starting material. Although removal of Triton X-100, to below its critical micelle concentration (cmc) is difficult, it was used for initial solubilisation as detergent exchange (to CHAPS) was achieved with the protein bound to the calmodulin affinity column.

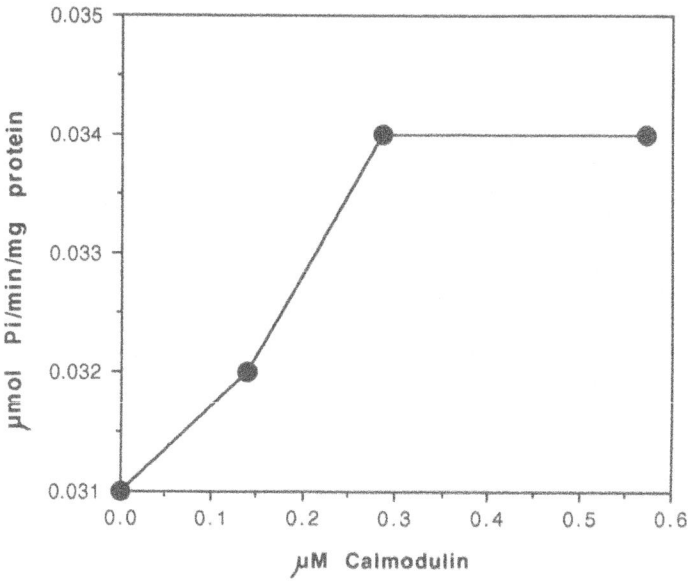

Figure 2 Calmodulin-stimulation of ATPase activity in Triton X-114 treated maize plasma membrane. ATPase activity was assayed in the integral protein (Triton X-114) phase obtained as described for Figure 1. ATP hydrolytic activity was assayed as described in 'Materials and Methods' in the presence of 0 - 0.6 μM bovine brain calmodulin.

Lipid was added to all column buffers and during solubilisation to preserve activity of the ATPase. After detergent exchange, the CaM ATPase was eluted using EGTA and stabilised with added calcium. This fraction (three parts) was added to buffer containing 26 mg/ml lipid (one part) to give a final CHAPS concentration of 10 mM. CHAPS was removed by dialysis over 60 hours with addition of Amberlite XAD2 to the final dialysis bath.

The achievement of a successful reconstitution is shown by the data presented in Figure 3. In the presence of calmodulin, calcium uptake increased with time, saturating after about 20 minutes. Initial rates of uptake, of about 10 nmol/min/mg protein compared favourably with an initial rate of ATP hydrolysis of around 2-6 μmol/min/mg given that no calcium chelator was present in the liposomes. Addition of the calcium ionophore A23187 dissipated the accumulated calcium from the liposomes to a level equivalent to basal, demonstrating that transport (as opposed to binding) was taking place. A23187 also stimulated ATP-hydrolytic

46

rates by about 60% indicating coupling of Ca^{2+} transport and ATP hydrolysis. However, this coupling is imperfect, due to the inherent leakiness of the system to calcium. It is also observed when other Ca^{2+} pumps are reconstituted.

Properties of the reconstituted system: The effect of added calmodulin on rates of ATP hydrolysis and calcium transport is shown in Figure 4. Both ATP hydrolysis and Ca^{2+} transport are clearly stimulated by calmodulin by about four fold over basal; there is some suggestion of saturatability of transport by around 0.1 μM calmodulin. Saturation of the red blood cell Ca^{2+} pump by calmodulin is calcium dependent and at calcium concentrations above 10 μM it is saturated by 0.1 μM calmodulin or less (Foder and Scharff, 1981); the assays described here were done in the presence cf around 40 μm free calcium.

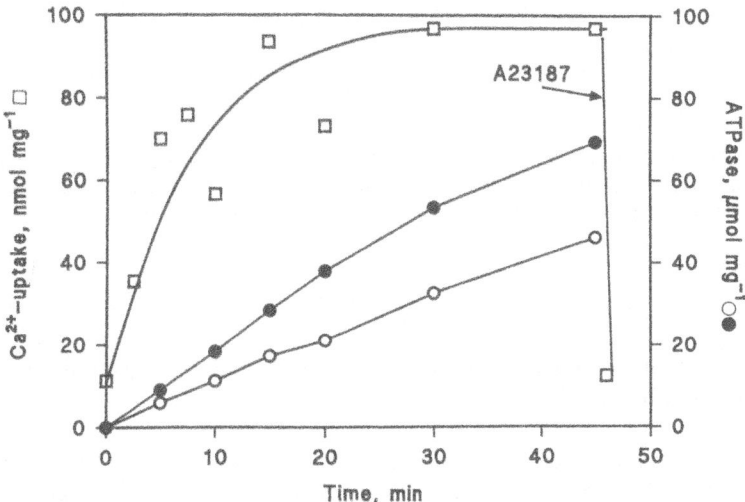

Figure 3 Time course of calcium uptake and ATP hydrolysis by reconstituted calmodulin-stimulated Ca^{2+} pump purified from cauliflower.
Calcium transport ([]) and ATP hydrolysis (●, ○) were assayed in the presence (closed symbols) or absence (open symbols) of ionophore A23187 (5 μM) and in the presence of 0.5 μM calmodulin. Assays were conducted as described in 'Materials and Methods'.

Inhibition of reconstituted CaM-ATPase by sodium *ortho*-vanadate indicated that the reconstituted calmodulin-stimulated activity was due to a P-type ion-translocating ATPase. Half-maximal activity occurred in the presence of 10 μM vanadate and activity was inhibited 85% with 200 μM vanadate. This compared favourably with the inhibition of activity shown previously for calmodulin-stimulated ATPase in maize membranes (Briars and Evans, unpublished). Activity was also inhibited by addition of erythrosin B (half-maximal inhibition = 10 μM; 90% inhibition by 100 μM). Calmodulin-stimulated NTPase activity showed some specificity for ATP (100%) as compared to GTP (50%), UTP and CTP (<10%). Optimum ATPase activity occurred at about pH 7.0. Preliminary investigation of the affinity of the calmodulin-stimulated ATPase for ATP demonstrated a bi-phasic response, with two K_m values of about 5 and 35 μM in the presence of calmodulin. Calmodulin-stimulated ATPase activity was dependent on Ca^{2+}.

47

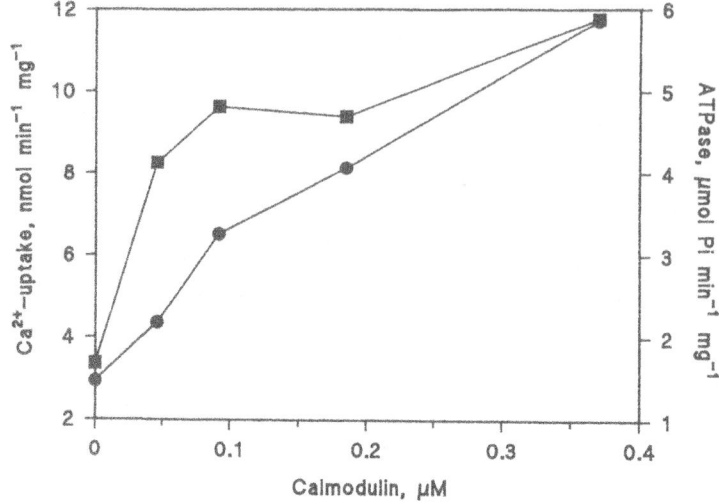

Figure 4 Calmodulin sensitivity of calcium transport and ATP-hydrolysis by reconstituted calmodulin-stimulated Ca^{2+} pump purified from cauliflower. Calcium transport (■) and ATP hydrolysis (●) were assayed as described in 'Materials and Methods' in the presence of 0 - 0.35 μM bovine brain calmodulin.

12. Phospholipid and Sterol Analysis of Affinity Purified Calmodulin-Stimulated ATPase from Maize

The phospholipid and sterol contents of the microsomal starting material and the material eluted by EGTA washing (and therefore co-purifying with the calmodulin-stimulated ATPase) are presented in Tables 3 and 4. The column wash (non-retained) fraction showed a slight percentage enrichment in PI and PS and a slight decrease in PC when compared with the membrane fraction. This finding paralleled the effect of Triton X-100 in solubilising microsomes (data not shown) and may therefore be due to detergent solubilisation effects. Similarly, the sterol profile also closely resembled that of the Triton X-100 solubilised membrane in being less depleted in cholesterol than the other sterols. However, the phospholipid and sterol content co-eluting with the calmodulin-stimulated ATPase was very different from the membrane of origin or the column wash fraction. All four phospholipids (PE, PI, PS and PC) were associated with the purified calmodulin-stimulated ATPase; PS and PI were enriched (as a percentage total phospholipid) as compared to the microsomal fraction, while PE and PC were depleted; percentage recoveries of all four from the starting material were between 2.5% and 3.5%. A far larger disparity between membrane and eluted fraction was apparent when sterols were analysed. A very high proportion of the cholesterol in the microsomal membrane co-eluted with the calmodulin-stimulated ATPase (62.5%), whilst the other three sterols were less abundant (together representing about 35% of the eluted fraction) and showed lower recoveries from the microsomal membrane (between 0.5% and 1.6%). Comparison of the bond indices of the fatty acids are presented in Table 5. The purified peak fraction showed a substantially lower bond-index (i.e., higher degree of saturation) than any of the membrane fractions analysed and than the non-retained fraction. This was the case for all four phospholipids analysed.

Table 3 Percentage composition of phospholipids in maize membranes and co-purifying with calmodulin (CaM)-stimulated ATPase from Maize.

Fraction	PE	PI	PS	PC
Microsomes	31	11	3	54
Non-retained fraction	31	16	9	43
Lipid co-eluting with CaM ATPase	26	32	19	25
% start material co-eluting with CaM ATPase	0.4	3.7	3.0	0.25

(Expressed as % PI + PS + PE + PC assayed).   Total (PI+PS+PC+PE) for:- Microsomes: 4.9mg; Non-retained fraction: 2.09 mg; lipid co-eluting with CaM
ATPase: 0.026 mg total.

Table 4 Percentage composition of sterols co-purifying with the calmodulin (CaM)-stimulated ATPase from Maize.

Fraction	Cholesterol	Campesterol	Stigmasterol	Sitosterol
Microsomes	1.9	19.6	47	32
Sterols co-eluting with CaM ATPase	55	13	11	21
% start material in CaM ATPase	62.5	1.6	0.5	1.6

Data expressed as % sterols assayed.   Totals (Cholesterol + sitosterol + campesterol + stigmasterol) for:- Microsomes: 0.811 mg; Non-retained Fraction: 1.03 mg; sterols co-purifying with CaM ATPase: 0.018 mg (total).

Table 5 Bond indices of fatty acids co-purifying with calmodulin (CaM)-stimulated ATPase from Maize.

Fraction	Mean (PE+PI+PS+PC)
Microsomes	132
Non-retained fraction	150
Lipid co-eluting with CaM ATPase	56

Bond index = % 16.1 + 18.1 + (2x %18.2) + (3x 18.3)

Where is the calmodulin-stimulated calcium pump?

It is now obviously unwise to make the assumption that the calmodulin-stimulated ATPase is exclusively a plasma membrane activity, however persuasive homology with mammalian calcium pumps may be (Briars, Kessler, and Evans, 1988; Briars, Dewey, and Evans, 1989). From our own work, a plasma membrane location for at least some of the activity remains a strong possibility. In addition to the data presented here showing calmodulin-stimulated ATP hydrolysis after removal of endogenous calmodulin, we have shown that plasma membrane purified by two-phase partition forms a phosphorylated intermediate of higher M_r than the proton pump which has properties consistent with it being a Ca^{2+} pump. Furthermore, gels of the EGTA eluent of calmodulin-affinity purifications of plasma membrane also show an SDS-PAGE profile containing a ca. 135,000 M_r polypeptide similar to that shown in microsomes to be the calmodulin-stimulated ATPase (Askerlund and Evans, unpublished). The small percentage (ca. 10%) stimulation of ATP-hydrolysis in Triton X-114 treated plasma membrane is probably all that might be expected given the presence of other, probably more abundant ATP-hydrolysers (e.g., the PM proton pump) which will be active in the assay conditions used.

So how can the apparent absence of calmodulin-stimulated calcium transport from purified plasma membrane be explained? The presence of large amounts of endogenous calmodulin may indicate that the pump is not regulated by changes in calmodulin concentrations *in vivo*, but is (at least in most conditions) fully saturated. Alternatively, calmodulin may become associated with the membrane during purification, and the pump may not be saturated with it *in vivo*. Given that the response of calmodulin binding proteins to calmodulin varies as a function of both the free calcium concentration and the calmodulin concentration, calmodulin remains as an important regulator in either case. It remains to be seen whether plasma-membrane associated calmodulin provides an explanation for the apparent absence of calmodulin-stimulation from purified plasma membrane observed by other workers (for example, Rasi-Caldogno, *et al.*, 1987; Kasai and Muto, 1990; Gräf and Weiler, 1989, 1990).

Our work, as yet, does not make it possible for us to conclude whether calmodulin-stimulated calcium transport is also a property of other membranes (as suggested by other workers, Table 2). However, it is worth noting that we appear to obtain far higher activities from "microsomes" than from plasma membrane, suggesting that another location is a possibility. In spinach beet for instance, purification from plasma membrane yields less than 2% of the activity obtained from cauliflower microsomes, while around 10% of the microsomal protein is likely to be plasma membrane. This strongly suggests that proteolysis, the presence of endogenous inhibitors, or saturation with calmodulin may mask activity in membrane fractions.

13.   Properties of the Reconstituted Calmodulin-Stimulated Calcium Pump

Successful functional reconstitution of the calmodulin-stimulated ATPase purified by affinity chromatography provides the first conclusive evidence that it is a calcium pump. The similarity between the properties described for it and those described for calmodulin-stimulated calcium transport in membranes suggest that this purified activity is the microsomal calcium pump previously described (Dieter and Marmé, 1980, 1981a,b, 1983). The reconstituted system shows a high degree of coupling between Ca transport and ATP hydrolysis given that no calcium chelators are present in the liposomes and a dose-response curve for calmodulin similar to that obtained for plant membranes and for red blood cells at comparable calcium concentrations (Foder and Scharff, 1987). Inhibition by vanadate and erythrosin B are consistent with a P-type calcium pump. The nucleotide specificity (NTPase activity) observed is similar to that for the mammalian calmodulin-stimulated ATPase excepting the high activity sustained by GTP (e.g., Rega, Richards, and Garrahan, 1973) and to that observed for basal (i.e., non-calmodulin-stimulated) Ca^{2+} transport in plant membranes, including purified plasma membrane (e.g., Gross and Marmé, 1978; Giannini, Ruis-Christin, and Briskin, 1987; Rasi-Caldogno, 1989; Gräf and Weiler, 1989). Both the pH optimum and biphasic kinetics for ATP strongly resemble those observed for the mammalian calmodulin-stimulated ATPase (Richards, Rega, and Garrahan, 1978). Indeed, while the subcellular location of the plant calmodulin-stimulated ATPase remains in some doubt, there is a striking resemblance between it and the mammalian calmodulin-stimulated ATPase which is generally a plasma membrane activity (see Carafoli, 1987).

14.   Lipid Environment

Investigations to date have revealed that large amounts of cholesterol co-purify with the calmodulin-stimulated ATPase, together with other sterols and phospholipids. Cholesterol is believed to reduce the back-flow of calcium through the SR calcium pump in reconstituted systems (A.G. Lee *personal communication*) and its presence may therefore have a significant effect on pump activity *in vivo*. Lipids co-eluting also appear to show a greater degree of saturation as well as slight enrichment of PS and PI. We are already involved in further investigation to discover whether the lipids co-purifying with the calmodulin-stimulated calcium pump are representative of its membrane environment.

Acknowledgements

We are grateful to the Agricultural and Food Research Council for their support for parts of this work and to Professor C. Larsson, Lund, Sweden, for his kind gift of polyethylene glycol 3350 and Dextran T500. David Evans is a Royal Society 1983 University Research Fellow and Per Askerlund is a Swedish NFR postdoctoral Research Fellow. Frederica Theodoulou is a Science and Engineering Research Council graduate student.

REFERENCES

ANDREEV, I.G., KORENK'KOV, V., and MOLOTOVSKY, Y.G., 1990. calmodulin-stimulation of nCa^{2+}/H^+ antiport across the vacuolar membrane of sugar-beet taproot. *Journal of Plant Physiology*, 136, 3-7.

ASKERLUND, P., LARSSON, C., and WIDELL, S., 1988. Localization of donor and acceptor sites of NADH dehydrogenase activities using inside-out and right-side out plasma membrane vesicles from plants. *FEBS Letters*, 239, 23-28.

ASKERLUND, P., LAURENT, P., NAKAGAWA, H., and KADER, J.-C., 1991. NADH - ferricyanide reductase of leaf plasma membranes. Partial purification and immunological relation to potato tuber microsomal NADH-ferricyanide reductase and spinach leaf NADH-nitrate reductase. *Plant Physiology*, 95, 6-13.

BLUMWALD, E., and Poole, R.J., 1986. Kinetics of Ca^{2+}/H^+ antiport in isolated tonoplast vesicles from storage tissue of *Beta vulgaris* L. *Plant Physiology*, 80, 727-731.

BRIARS, S.A., DEWEY, F.M., and EVANS, D.E., 1989. The calcium pumping ATPase of the plant plasma membrane. In *Plant membrane transport, the current position*. Eds J. Dainty, M.I.DeMichelis, E. Marré, F. Rasi-Caldogno, Elsevier, Amsterdam. pp 231-236.

BRIARS, S.A., and EVANS, D.E., 1989. The calmodulin-stimulated ATPase of maize coleoptiles forms a phosphorylated intermediate. *Biochemical and Biophysical Research Communications*, **159**, 185-191.

BRIARS, S.A., KESSLER, F., and EVANS, D.E., 1988. The calmodulin-stimulated ATPase of maize coleoptiles is a 140,000-M_r polypeptide. *Planta*, **176**, 283-285.

BRISKIN, D.P., 1990. Ca^{2+}- translocating ATPase of the plant plasma membrane. *Plant Physiology*, **94**, 397-400.

BRAUER, D., SCHUBERT, C., and TSU, S.-I., 1990. Characterisation of a Ca^{2+}-translocating ATPase from corn root microsomes. *Physiologia Plantarum*, **78**, 335-344.

BUSH, D.S., BISWAS, A.K., and JONES, R.L., 1989. Gibberellic-acid-stimulated Ca^{2+} accumulation in endophasmic reticulum of barley aleurone: Ca^{2+} transport and steady-state levels. *Planta*, **178**, 411-420.

BUSH, D.R., and SZE, H., 1986. Ca^{2+}-transport in tonoplast and endoplasmic reticulum vesicles isolated from cultured carrot cells. *Plant Physiology*, **80**, 549-55.

BUTCHER, R.D., and EVANS, D.E., 1987a. Calcium transport by pea root membranes. I. Purification of membranes and characteristics of uptake. *Planta*, **172**, 265-278.

BUTCHER, R.D., and EVANS, D.E., 1987b. Calcium transport by pea root membranes. II. Effects of calmodulin and inhibitors. *Ibid*, **172**, 273-279.

CARAFOLI, E., 1987. Intracellular calcium homeostasis. *Annual Review of Biochemistry*, **56**, 395-333.

COLLINGE, M., and TREWAVAS, A.J., 1989. The location of calmodulin in the pea plasma membrane. *Journal of Biological Chemistry*, **15**, 8865-8872.

COOKE, D.T., BURDEN, S., CLARKSON, D.T., and JAMES, C., 1989. Xenobiotic induced changes in membrane lipid composition: effects on plasma membrane ATPases. *British Plant Regulator Group Monograph* 18, Mechanisms of Transport Processes. pp 41-53.

DIETER, P., and MARMÉ, D., 1980. Ca^{2+} -transport in mitochondrial and microsomal fractions from higher plants. *Planta*, **150**, 1-8.

DIETER, P., and MARMÉ, D., 1981a. Far-red light irradiation of intact corn seedlings affects mitochondrial and calmodulin-dependent microsomal Ca^{2+} transport. *Biochemical and Biophysical Research Communications*, **101**, 749-755.

DIETER, P., and MARMÉ, D., 1981b. A calmodulin-dependent, microsomal ATPase from corn (*Zea mays*). *FEBS Letters*, **125**, 245-248.

DIETER, P., and MARMÉ, D., 1983. The effect of calmodulin and far-red light on the kinetic properties of the mitochondrial Ca^{2+} transport system from corn. *Planta*, **159**, 277-281.

EVANS, D.E., 1988. Regulation of cytoplasmic free calcium by plant cell membranes. *Cell Biology International Reports*, **12**, 383-396.

EVANS, D.E., BRIARS, S.-A., and WILLIAMS, L.E., 1991. Active calcium transport by plant cell membranes. *Journal of Experimental Botany*, **42**, 285-303.

FODER, B. and SCHARFF, O., 1981. Decrease of apparent calmodulin affinity of erythrocyte $(Ca^{2+}-Mg^{2+})$-ATPase at low Ca^{2+} concentrations. *Biochimica et Biophysica Acta*, **649**, 367-371.

FREDRIKSON, K., and LARSSON, C., 1989. Activation of $1,3\beta$-glucan synthase by Ca^{2+}, spermine and celloboise. - Localisation of active sites using inside-out plasma membrane vesicles. *Physiologia Plantarum*, **77**, 196-201.

GRÄF, P., and WEILER, E.W., 1989. ATPase-driven transport in sealed plasma membrane vesicles prepared by aqueous two-phase partitioning from leaves of *Commelina communis*. *Physiologia Plantarum*, **75**, 469-478.

GRÄF, P., and WEILER, E.W., 1990. Functional reconstitution of an ATP-driven Ca^{2+}-transport system from the plasma membrane of *Commelina communis*. *Physiologia Plantarum*, **75**, 469-478.

GIANNINI, J.L., RUIS-CHRISTIN, J., and BRISKIN, D.P., 1987. Calcium transport in sealed vesicles from red beet (*Beta vulgaris* L.) storage tissue. II. Characterisation of $^{45}Ca^{2+}$ uptake into plasma membrane vesicles. *Plant Physiology*, **85**, 1137-1142.

GROSS, J. and MARMÉ, D., 1978. ATP-dependent Ca^{2+}-uptake into plant membrane vesicles. *Proceedings of the National Academy of Sciences, USA*, **75**, 1232-1236.

HELMKE, S.M. and HOWARD, B.D., 1987. Fractionation and solubilisation of the Sarcoplasmic Reticulum Ca^{2+} pump solubilised and stabilised by CHAPS/lipid micelles. *Membrane Biochemistry*, **7**, 1-22.

KANNANGARA, C.G., GOUGH, S.P., HANSEN, B., RASMUSSEN, Y. and SIMPSON, D.J., 1977. A homogenizer with replacement razor blades for bulk isolation of active barley plastids. *Carlsberg Research Communications*, **42**, 431-439.

KASAI, M. and MUTO, S., 1991. Solubilisation and reconstitution of Ca^{2+} pump from corn leaf plasma membrane. *Plant Physiology*, **96**, 565-570.

KJELLBOM, P., LARSSON, C., ROCHESTER, C., and ANDERSSON, B., 1989. Integral and peripheral proteins of the spinach leaf plasma membrane. *Plant Physiology and Biochemistry*, **27**, 169-174.

LARSSON, C., WIDELL, S., and KJELLBOM, P., 1987. Preparation of high purity plasma membranes. *Methods in Enzymology*, **148**, 558-568.

MALATIALY, L., GREPPIN, H. and PENEL, C., 1988. Ca^{2+} uptake by tonoplast and plasma membrane vesicles from spinach leaves. *FEBS Letters*, **233**, 196-200.

MEIR, P., JUNIPER, B.E., and EVANS, D.E., 1991. Regulation of free calcium concentration in the pitchers of the carnivorous plant *Sarracenia purpurea*; a model for calcium in the higher plant apoplast? *Annals of Botany*, in the press.

NIGGLI, V., PENNISTON, J.T., and CARAFOLI, E., 1979. Purification of the $(Ca^{2+}+Mg^{2+})$-ATPase from human erythrocyte membranes using a calmodulin-affinity column. *Journal of Biological Chemistry*, **254**, 9955-9958.

RASI-CALDOGNO, F., PUGLIARELLO, M.C. and DEMICHELIS, M.I., 1987. The Ca^{2+}-transport ATPase of plant plasma membrane catalyses a nH^+/Ca^{2+} exchange. *Plant Physiology*, **83**, 994-1000.

RASI-CALDOGNO, F., OLIVARI, C., and DeMICHELIS, M.I., 1989. Identification and characterisation of the Ca^{2+}-ATPase which drives active transport of Ca^{2+} at the plasma membrane of radish seedlings. *Plant Physiology*, **90**, 1429-1434.

REGA, A.F., RICHARDS, D.E., and GARRAHAN, P.J., 1973. Calcium ion-dependent *p*-nitrophenylphosphatase activity and calcium ion dependent adenosine triphosphatase activity from human erythrocyte membranes. *Biochemical Journal*, **136**, 185-194.

RICHARDS, D.E., REGA, A.F., and GARRAHAN, P.J., 1978. Two classes of site for ATP in the Ca^{2+} ATPase from human red cell membranes. *Biochimica et Biophysica Acta*, **511**, 194-201.

ROBINSON, C., LARSSON, C., and BUCKHOUT, T.K., 1988. Identification of a calmodulin-stimulated $(Ca^{2+} + Mg^{2+})$-ATPase in a plasma membrane fraction isolated from maize (*Zea mays* L.) leaves. *Physiologia Plantarum*, **72**, 177-184.

SCHUMAKER, K.S., and SZE, H., 1985. A Ca^{2+}/H^+ antiport system driven by the proton electrochemical gradient of a tonoplast H^+-ATPase from oat roots. *Plant Physiology*, **79**, 1111-1117.

STOSCHECK, C.M., 1990. Quantitation of protein. *Methods in Enzymology*, **182**, 50-68.

WIDELL, S. and LARSSON, C., 1990. A critical evaluation of markers used in plasma membrane purification. In *The plant plasma membrane, structure function and molecular biology*. Eds C. Larsson, and I.M. Møller. Springer Verlag, Berlin. pp 16-43.

WILLIAMS, L.E., SCHUELER, S.B. and BRISKIN, D.R., 1990. Further characterisation of the red-beet plasma membrane Ca^{2+}-ATPase using GTP as an alternative substrate. *Plant Physiology*, **92**, 747-754.

ZOCCHI, G., 1988. Separation of membrane vesicles from maize roots having different calcium transport activities. *Plant Science Letters*, **54**, 103-107.

The Plasma Membrane Calcium Pump: Studies on Membrane Vesicles Isolated from Higher Plants

M.I. De Michelis[1], F. Rasi-Caldogno[2], M.C. Pugliarello[2], C. Olivari[2] and A. Carnelli[2]

1. Instituto di Botanica
    Università de Messina, C.P. 58
    98166 Messina-S
    Agata, Italy

2. Centro di Studio del C.N.R.
    sulla Biologia Cellulare e Molecolare delle Piante
    Departimento di Biologia
    Sessione di Fisiologia e Biochimica delle Piante
    Via Celoria 26
    20133, Milan, Italy

1. Introduction

Free Ca^{2+} concentration in the cytoplasm of living plant cells is kept in the submicromolar range against a steep electrochemical gradient which would favour Ca^{2+} influx into the cytoplasm both from intracellular compartments and from the apoplast (Poovaiah and Reddy, 1987; Felle 1989; Miller, Vogg and Sanders, 1990). A variety of hormonal and environmental stimuli induce an increase in cytoplasmic free Ca^{2+} concentration, most likely via the opening of regulated Ca^{2+} channels in endomembranes and/or in the plasma membrane (PM) (Poovaiah and Reddy, 1987; Tester, 1990).

Re-establishment of cytoplasmic Ca^{2+} homeostasis relies on the activity of active systems of Ca^{2+} transport which catalyse the extrusion of Ca^{2+} from the cytoplasm to intracellular compartments and/or to the apoplast (Evans, Briars and Williams, 1991 and references therein). Since intracellular compartments have limited capacities, maintenance of Ca^{2+} homeostasis must involve, at least on a long term basis, Ca^{2+} extrusion to the apoplast across the PM (Miller *et al.*, 1990; Evans *et al.*, 1991).

During the last few years, the availability of transport competent PM vesicles has enabled researchers to show that Ca^{2+} uptake into inside-out PM vesicles can be energised by ATP. The bulk of ATP-dependent Ca^{2+} uptake is insensitive to protonophores such as

Transport and Receptor Proteins of Plant Membranes
Edited by D.T. Cooke and D.T. Clarkson, Plenum Press, New York, 1992

55

FCCP and nigericin or to ammonia, thus indicating that it is driven by a primary ATP-fuelled Ca^{2+} pump (reviewed in Briskin 1990; De Michelis, Rasi-Caldogno and Pugliarello, 1991, Evans et al., 1991). Recently it has been proposed that a secondary active system of Ca^{2+} transport driven by the proton gradient generated by the PM H^+-ATPase is present also in PM from Zea mays L. leaves (Kasai and Muto, 1990). However, this seems to be an exception to the normal systems observed in PM from other plant materials, so only the primary ATP-fuelled Ca^{2+} pump of the PM will be dealt with here.

The ATP-fuelled mechanism of Ca^{2+} transport has been described in some detail in PM vesicles from various plant materials (Butcher and Evans, 1987a,b; Giannini, Ruiz-Cristin, and Briskin, 1987; Rasi-Caldogno, Pugliarello, and De Michelis, 1987; Malatialy, Greppin, and Penel, 1988; Zocchi, 1988; Rasi-Caldogno, Pugliarello, Olivari, and De Michelis, 1989a,b; Gräf and Weiler, 1989, Kasai and Muto, 1990; Williams, Schueler, and Briskin, 1990) and also in proteoliposomes reconstituted with solubilised PM proteins (Gräf and Weiler, 1990). It has a broad, slightly alkaline pH optimum (between pH 7.0 and 7.5), is strictly Mg^{2+}-dependent and can utilise ATP and also ITP or GTP as substrates, the rate of Ca^{2+} uptake sustained by ITP or GTP being about 50 to 70% that sustained by ATP. ATP-dependent Ca^{2+} uptake has a very high apparent affinity for ATP, the K_m value being about 10 to 20 μM, and is half-saturated by Ca^{2+} in the μM concentration range (for a review see Briskin 1990a, De Michelis et al., 1991, Evans et al., 1991). Among the inhibitors tested, the most interesting is the iodinated derivative of fluorescein, erythrosin B (EB), which shows a high degree of specificity for the PM Ca^{2+} pump, as compared to other membrane-bound ATPases of plant cells: ATP-dependent Ca^{2+} uptake into PM vesicles is 50% inhibited by 0.01 to 0.1 μM EB, and abolished by 1 μM EB (Gräf and Weiler, 1989, 1990; Rasi-Caldogno et al., 1987, 1989a; Williams et al., 1990), a concentration which has very little effect on the other ATPases tested (Cocucci, 1986; Rasi-Caldogno et al., 1987).

In this paper we show that, by exploiting the different biochemical characteristics of the PM Ca^{2+} pump and of the PM H^+-ATPase - which is the most active ATPase in the plant PM - it is also possible to measure the hydrolytic activity of the PM Ca^{2+} pump in native PM vesicles, as the Ca^{2+}-dependent portion of the ATPase or ITPase activity. We then discuss the dependence of the PM Ca^{2+} pump activity on Ca^{2+} concentration and pH, as well as its regulation by the calcium binding protein calmodulin and the mechanism through which it drives Ca^{2+} transport.

2. Measurements of the Hydrolytic Activity of the PM Ca^{2+} Pump

Demonstrating the hydrolytic activity of the PM Ca^{2+} pump in native PM has proven difficult, due to the simultaneous operation of the much more abundant H^+-ATPase, which is inhibited by μM concentrations of Ca^{2+} in the same range required to activate the Ca^{2+}-ATPase (Briskin, 1990b; Cocucci and Marrè, 1984).

So, when the ATPase activity of the PM from radish (Raphanus sativus L.) is assayed at the optimum pH for the activity of the PM H^+-ATPase (i.e. pH 6.6), addition of 30 μM Ca^{2+} results in a decrease of ATPase activity. Therefore, the operation of a Ca^{2+}-dependent ATPase activity can be inferred only from the observation that the inhibitory effect of Ca^{2+} is much stronger in the presence of μM EB than in its absence (Table 1).

However, a Ca^{2+}-dependent ATPase activity can be demonstrated by exploiting the different pH profiles of the two enzymes. At pH 7.5, where the activity of the PM Ca^{2+} pump is still about optimal, but that of the PM H^+-ATPase is markedly lower than at pH

Table 1 Effect of Ca^{2+} and erythrosin B (EB) on the PM ATPase activity of radish. Data are from Rasi-Caldogno *et al.*, (1989a) with modifications. Calcium was added as 30 μM CaCl$_2$ in the absence of ETGA; EB was 1 μM and A23187, 2 μM.

| | ATPase activity (μmol Pi mg^{-1} prot h^{-1}) | | |
	EGTA	Ca^{2+}	Ca^{2+}-dependent
pH 6.6			
A23187	6.0	5.7	- 0.3
" + EB	5.6	4.5	- 1.1
pH 7.5			
none	2.3	2.7	+ 0.4
A23187	2.3	3.1	+ 0.8
" + EB	2.0	1.9	- 0.1

6.6 (Briskin 1990b, Cocucci and Marré, 1984), addition of 30 μM Ca^{2+} increases the ATPase activity. This Ca^{2+}-dependent ATPase activity is stimulated by the calcium ionophore A23187, as expected for an enzyme driving Ca^{2+} accumulation into the vesicles and is fully inhibited by μM EB (Table 1, lower part).

Working at pH 7.5, we have shown that the characteristics of the Ca^{2+}-dependent ATPase activity are very similar to those of ATP-dependent Ca^{2+} uptake (Rasi-Caldogno *et al.*, 1989a, b). Also, the functional molecular weights of ATP-dependent Ca^{2+} uptake and of the Ca^{2+}-dependent ATPase activity, measured with the radiation inactivation technique, are very similar (Rasi-Caldogno, Pugliarello, Olivari, De Michelis, Gambarini, Colombo and Tosi, 1990). These results altogether indicate that the two activities represent respectively the transport and the hydrolytic activity of the same enzyme, the PM Ca^{2+}-ATPase.

However, measuring the hydrolytic activity of the PM Ca^{2+}-ATPase using ATP as a substrate imposes some strict limitations and can be difficult, since even at pH 7.5 the PM H^{+}-ATPase contributes substantially to the total activity measured and any treatment can be expected to influence both activities in the same or in opposite directions.

We circumvented this problem by taking advantage of another major difference between the PM H^{+}-ATPase and the PM Ca^{2+} pump, namely, that whilst the PM H^{+}-ATPase is highly specific for ATP as a substrate (Briskin 1990b, and references therein), the PM Ca^{2+} pump can use also ITP or GTP (Butcher and Evans, 1987a, Giannini 1987, Gräf and Weiler, 1989, Rasi-Caldogno *et al.*, 1989a, Williams *et al.*, 1990). Table 2 shows that in PM from radish the bulk of ITPase activity is Ca^{2+}- and Mg^{2+}-dependent. The Ca^{2+}-dependent ITPase activity is stimulated by A23187, inhibited by μM EB and has the same biochemical characteristics of ATP-dependent Ca^{2+} uptake (unpublished data from the authors' laboratories).

The availability of a simple, reliable method to measure the hydrolytic activity of the PM Ca^{2+}-ATPase in native PM vesicles will prove useful in further studies of this enzyme, in particular with respect to its interactions with other components of the PM. As an example, analysis of the effect of detergents on the activity of the PM Ca^{2+}-ATPase in PM vesicles from radish, isolated by the phase-partitioning technique which is thought to produce predominantly right-side-out PM vesicles, shows that the PM Ca^{2+}-ATPase activity is highly sensitive to detergents. Only the very mild polyoxyethylene ether, Brij 58, increases the Ca^{2+}-dependent ITPase activity. With lysolecithin or Triton X-100, the unmasking effect of the detergents is largely or completely masked by their inhibitory effect on Ca^{2+}-dependent ITPase activity (unpublished data from the authors' laboratories).

Table 2 Effect of Ca^{2+} on the ITPase activity of PM from radish.
The ITPase activity of a purified PM fraction was assayed at pH 6.9 in a standard ATPase assay medium containing 1 mM EGTA plus or minus $CaCl_2$ to give 100 μM free Ca^{2+}; A23187 was supplied at 5 μM and erythrosin B (EB) at 1 μM.

| | ATPase activity (μmol Pi mg^{-1} prot h^{-1}) | | |
	Ca^{2+}	+ Ca^{2+}	Ca^{2+}-dependent
Control	0.23	0.76	0.53
- Mg^{2+}	0.07	0.08	0.01
+ A23187	0.22	1.03	0.81
" + EB	0.23	0.26	0.03

Indeed, the choice of the detergent has proved critical in the solubilisation of functional enzyme from the PM (Gräf and Weiler, 1990; Kasai and Muto, 1991). These results suggest that membrane phospholipids closely associated with the enzyme may play a crucial role in the activity of the PM Ca^{2+}-ATPase.

Finally, we want to stress that the possibility of using ITP or GTP (Williams *et al.*, 1990) as alternative substrates for assaying the PM Ca^{2+}-ATPase activity is useful for studying not only the hydrolytic activity of the enzyme, but also its transport activity, because it is presently the only available means to study the activity of the PM Ca^{2+}-ATPase in native PM under conditions in which the H^+-ATPase is not operating.

3.   Dependence of the Ca^{2+}-ATPase Activity on the Concentration of Ca^{2+}

Our first measurements of the dependence of the PM Ca^{2+}-ATPase activity of radish on the concentration of free Ca^{2+}, performed in EGTA buffered media, gave an apparent K_m value of about 70 nM (Rasi-Caldogno *et al.*, 1989a,b). This was much lower than those measured, in the absence of EGTA, for the PM Ca^{2+}-pump from other plant materials, which ranged between 2 and 8 μM (Giannini *et al.*, 1988; Gräf and Weiler, 1989, 1990; Williams *et al.*, 1990). However, free Ca^{2+} concentrations reported in Rasi-Caldogno *et al.*, 1989a,b were wrong, because they were computed on the basis of a value of the association constant of the Ca-EGTA buffers with a Ca^{2+}-selective electrode. Using the experimentally determined value of the association constant to calculate free Ca^{2+} concentrations, we showed that in radish the apparent K_m of the PM Ca^{2+}-ATPase is also in the μM range. The dependence on free Ca^{2+} concentration is very similar for the hydrolytic and the transport activity of the enzyme and is the same when ATP or ITP are used as substrates. The apparent K_m for Ca^{2+} is similar in the presence and in the absence of EGTA, indicating that EGTA has no major effect on the activity of the PM Ca^{2+}-ATPase of plants (unpublished data from the authors' laboratories).

Table 3 shows that the apparent K_m for Ca^{2+} of the PM Ca^{2+}-ATPase is strongly influenced by the pH of the assay medium; lowering the pH from 7.5 to 6.9 has little effect on V_{max} of the enzyme but reveals an increase of its apparent K_m for Ca^{2+} from 2μM to 12 μM.

Table 3 Effect of pH on the kinetic parameters of the PM Ca^{2+}-ATPase of radish.
V_{max} and K_m values were estimated from the dependence on free Ca^{2+} concentration of Ca^{2+}-dependent ITPase activity in a purified PM fraction. Free Ca^{2+} was buffered with 1-2 mM EGTA; free Ca^{2+} concentrations were computed using experimentally determined values of the apparent association constant of the Ca-EGTA complex (5.0 x 10^5 M^{-1} at pH 6.9 and 7.6 x 10^6 M^{-1} at pH 7.5)

	Ca^{2+}-dependent ITPase	
	V_{max} (μmol Pi mg^{-1} prot h^{-1})	Apparent K_m (μM)
pH 6.9	1.6	12
pH 7.5	1.9	3

The fact that the PM Ca^{2+}-ATPase is half-saturated by Ca^{2+} concentrations in the μM range has important implications with respect to its potential physiological role, since it means that the enzyme is able to respond to the increases of cytoplasmic free Ca^{2+} concentrations observed under physiological conditions by greatly enhancing the rate of Ca^{2+} extrusion through the PM. Also, the variations of its apparent K_m with pH, observed in the physiological range of cytoplasmic pH, can have important physiological implications, being a possible link between pH and free Ca^{2+} homeostasis in the cytoplasm (Felle, 1988).

4. Regulation of the PM Ca^{2+}-ATPase Activity by Calmodulin

The plant PM Ca^{2+}-ATPase has a number of structural and biochemical similarities with the erythrocyte enzyme (Carafoli, 1991; Evans *et al.*, 1991; De Michelis *et al.*, 1991). The erythrocyte Ca^{2+}-ATPase is stimulated by calmodulin (CaM), as is the plant enzyme (Malatialy *et al.*, 1988; Zocchi, 1988; Robinson, Larsson and Buckout, 1988; Williams *et al.*, 1990), although in some PM preparation and/or under some experimental conditions no effect of exogenous CaM is observed (Butcher and Evans, 1987b; Gräf and Weiler, 1989, 1990; Rasi-Caldogno *et al.*, 1989a; Kasai and Muto, 1991).

In PM from radish, exogenous CaM has very little effect on the Ca^{2+}-ATPase activity assayed at saturating Ca^{2+} concentrations, but it markedly stimulates the activity measured in the presence of sub-optimal concentrations of Ca^{2+}, both at pH 6.9 and at pH 7.5 (Table 4). The effect of CaM is abolished by μM EB and it is similar both in extent and in pattern when the transport or the hydrolytic activity of the PM Ca^{2+}-ATPase is measured and when ATP or ITP is used as a substrate.

Table 4 Effect of Calmodulin (CaM) on the PM Ca^{2+}-ATPase at pH 6.9 and 7.5.
The Ca^{2+}-dependent ITPase activity of radish PM was measured in EGTA-buffered media, in the presence and absence of 20 μg x ml^{-1} of bovine brain CaM.

Conditions	Control	Ca^{2+}-dependent ITPase + CaM (μmol Pi mg^{-1} prot h^{-1})	% stimulation
pH 6.9			
3 μM free Ca^{2+}	0.25	0.42	+ 68%
100 μM free Ca^{2+}	1.36	1.47	+ 8%
pH 7.5			
3 μM free Ca^{2+}	0.79	1.20	+ 52%
50 μm free Ca^{2+}	1.60	1.73	+ 8%

Analysis of the dependence of basal and CaM-stimulated activity on the concentration of free Ca^{2+} shows that activation by CaM determines a reduction of the apparent K_m for Ca^{2+} of the PM Ca^{2+}-ATPase to about half the value measured in its absence (unpublished results from the authors' laboratories).

In the light of these observations the variability of results obtained in different laboratories on PM isolated from various plant materials can be ascribed, at least in part, to the different experimental conditions.

The CaM-induced increase of apparent affinity of the PM Ca^{2+}-ATPase for Ca^{2+} is a further point of similarity between the plant and the erythrocyte enzyme (Carafoli, 1991). The major difference is the extent of stimulation by exogenous CaM. The effect we measured on the plant enzyme is relatively small compared with that measured on the erythrocyte enzyme, which can undergo a CaM-induced decrease of apparent K_m for Ca^{2+}

of up to 30-fold, accompanied by a substantial increase in V_{max} (Carafoli, 1991). However, it should be noted that in the absence of CaM the apparent K_m for Ca^{2+} of the plant enzyme is markedly lower than that of the erythrocyte enzyme, while in the presence of CaM the two values become much more similar (Carafoli, 1991). Moreover, CaM antagonists such as calmidazolium or W7 inhibit the radish PM Ca^{2+}-ATPase activity also in the absence of exogenous CaM (unpublished results from the authors' laboratories). These observations suggest that the Ca^{2+}-ATPase in PM isolated from radish may be in a particularly activated state. The PM Ca^{2+}-ATPase of *Beta vulgaris* L. failed to respond to exogenous CaM unless the membranes were extensively washed with EGTA before the assays (Williams *et al.*, 1990). Therefore, the PM from radish was extracted and washed in EGTA-containing media. However, further washes with EGTA, even in the presence of detergent, failed to improve the response to exogenous CaM. Thus, if the PM Ca^{2+}-ATPase from radish is activated by endogenous CaM, it must be CaM bound to the membrane in a Ca^{2+}-independent manner, not washable with EGTA (Collinge and Trewavas, 1989).

The high sensitivity of the PM Ca^{2+}-ATPase to detergents (see previous section) suggests that the enzyme might be activated by the lipid environment, which has been shown to modulate the activity of the erythrocyte Ca^{2+}-ATPase (Carafoli, 1991).

The CaM-induced increase in the apparent affinity of the PM Ca^{2+}-ATPase for Ca^{2+} may constitute an important factor in the control of cytoplasmic Ca^{2+} homeostasis, since it provides an amplification of the response of the PM Ca^{2+}-ATPase to an increase of cytoplasmic free Ca^{2+} concentration.

5. Mechanism of Ca^{2+} Transport of the PM Ca^{2+}-ATPase

The electrochemical gradient of Ca^{2+} across the PM of plant cells can be as high as 60 KJoules (Miller *et al.*, 1990). So, an ATPase catalysing a simple uniport of Ca^{2+} would not be thermodynamically competent to operate Ca^{2+} extrusion from the cytoplasm to the apoplast under physiological conditions. The other major ion gradient across the PM of plant cells is the H^+ gradient generated and maintained by the activity of the PM H^+-ATPase. We have analysed the effects of Ca^{2+} on the ATP-dependent H^+ fluxes in PM vesicles from radish. Addition of Ca^{2+} to the incubation medium markedly lowers the rate of ATP-dependent intravesicular acidification driven by the PM H^+-ATPase and increases the hyperpolarisation of the membrane potential. Micromolar EB, which only slightly inhibits ΔpH generation and membrane hyperpolarisation in the absence of Ca^{2+}, substantially suppresses the effects of Ca^{2+} on both parameters (Table 5). Moreover, a rapid efflux of H^+ is observed when, after reaching a steady state, ΔpH in the presence of MgATP, mM Ca^{2+} is added to block the PM H^+-ATPase and initiate Ca^{2+} uptake driven by the PM Ca^{2+}-ATPase. This efflux of H^+ is markedly slowed down by μM EB (Rasi-Caldogno *et al.*, 1987).

These results indicate that the PM Ca^{2+}-ATPase of plant cells, drives an exchange between Ca^{2+} and H^+ in a manner similar to the erythrocyte enzyme (Carafoli, 1991) and thus expels Ca^{2+} from the cytoplasm to the apoplast, using the energy provided by the hydrolysis of ATP, as well as the energy of the H^+ gradient built up by the PM H^+-ATPase.

The stoichiometry of the nH^+/Ca^{2+} exchange driven by the PM Ca^{2+}-ATPase of plant cells is as yet unknown. However, thermodynamic considerations based on the effect of external pH on cytoplasmic Ca^{2+} homeostasis in *Neurospora* suggest that it may operate with a stoichiometry of at least two H^+ per Ca^{2+} (Miller *et al.*, 1990).

Table 5 Effect of Ca^{2+} transport on the electrochemical proton gradient built up by the PM H^+-ATPase of radish.

Data are from Rasi-Caldogno *et al.*, (1987) with modifications. Assays were performed at pH 6.6 in the presence of 0.15 M KBr. Generation of ΔpH was measured as the initial rate of decrease of absorbance of Acridine orange; generation of the membrane potential difference ($\Delta\psi$) was measured as the maximum increase of absorbance of oxonol VI.

Conditions	MgATP-dependent ΔpH generation	MgATP-dependent $\Delta\psi$ generation
	relative activities	
1 mM EGTA	100	100
" + 1 μM EB	91	96
30 μM Ca^{2+}	34	177
" + 1 μM EB	72	90

REFERENCES

BRISKIN, D.P., 1990a. Ca^{2+}-translocating ATPase of the plant plasma membrane. *Plant Physiology*, **94**, 397-400.

BRISKIN, D.P., 1990b. The plasma membrane H^+-ATPase of higher plant cells: biochemistry and transport function. *Biochimica et Biophysica Acta*, **1019**, 95-109.

BUTCHER, R.D., and EVANS, D.E., 1987a. Calcium transport by pea root membranes. I. Purification of membranes and characteristics of uptake. *Planta*, **172**, 265-272.

BUTCHER, R.D., and EVANS, D.E., 1987b. Calcium transport by pea root membranes. II. Effects of calmodulin and inhibitors. *Planta*, **172**, 273-279.

CARAFOLI, E., 1991. Calcium pump of the plasma membrane. *Physiological Reviews*, **71**, 129-153.

COCUCCI, M.C., 1986. Inhibition of plasma membrane and tonoplast ATPases by erythrosin B. *Plant Science Letters*, **47**, 21-27.

COCUCCI, M.C., and MARRÈ, E., 1984. Lysophosphatidylcholine-activated, vanadate-inhibited, Mg^{2+}-ATPase from radish microsomes. *Biochimica et Biophysica Acta*, **771**, 42-52.

COLLINGE, M., and TREWAVAS, A.J., 1989. The location of calmodulin in the pea plasma membrane. *Journal of Biological Chemistry*, **264**, 8865-8872.

De MICHELIS, M.I., RASI-CALDOGNO, F., and PUGLIARELLO, M.C., 1991. The plasma membrane Ca^{2+} pump: potential role in Ca^{2+} homeostasis. In *Plant Growth Substances 1991*. Kluwer Academic Publishers, Dordrecht, The Netherlands. In press.

EVANS, D.E., BRIARS, S.A., and WILLIAMS, L.E., 1991. Active calcium transport by plant cell membranes. *Journal of Experimental Botany*, **236**, 285-303.

FELLE, H., 1988. Auxin causes oscillations of cytosolic free calcium and pH in *Zea mays* coleoptiles. *Planta*, **174**, 495-499.

FELLE, H., 1989. Ca^{2+}-selective microelectrodes and their application to plant cells and tissues. *Plant Physiology*, **91**, 1239-42.

GIANNINI, J.L., RUIZ-CRISTIN, J., and BRISKIN, D.P., 1987. Calcium transport in sealed vesicles from red beet (*Beta vulgaris* L.) storage tissue. II. Characterisation of $^{45}Ca^{2+}$ uptake into plasma membrane vesicles. *Plant Physiology*, **85**, 1137-1142.

GRÄF P., and WEILER, E.W., 1989. ATP-driven Ca^{2+} transport in sealed plasma membrane vesicles prepared by aqueous two-phase partitioning from leaves of *Commelina communis*. *Physiologia Plantarum*, **75**, 469-478.

GRÄF P., and WEILER, E.W., 1990. Functional reconstitution of an ATP-driven Ca^{2+}-transport system from the plasma membrane of *Commelina communis* L. *Plant Physiology*, **94**, 634-640.

KASAI, M., and MUTO, S., 1991. Ca^{2+} pump and Ca^{2+}/H^+ antiporter in plasma membrane vesicles isolated by aqueous two-phase partitioning from corn leaves. *Journal of Membrane Biology*, **114**, 133-142.

KASAI, M., and MUTO, S., 1990. Solubilisation and reconstitution of Ca^{2+} pump from corn leaf plasma membrane. *Plant Physiology*, **96**, 565-570.

MALATIALY, L., GREPPIN, H., and PENEL, C., 1988. Calcium uptake by tonoplast and plasma membrane vesicles from spinach leaves. *FEBS Letters*, **233**, 196-200.

MILLER, A.J., VOGG, G., and SANDERS, D., 1990. Cytosolic calcium homeostasis in fungi: roles of plasma membrane transport and intracellular sequestration of calcium. *Proceedings of the National Academy of Science, U.S.A.*, **87**, 9348-9352.

POOVAIAH, B.W., and REDDY, A.S.N., 1987. Calcium messenger system in plants. *CRC Critical Reviews in Plant Sciences*, **6**, 47-103.

RASI-CALDOGNO, F., PUGLIARELLO, M.C., and DE MICHELIS, M.I., 1987. The Ca^{2+} transport ATPase of plant plasma membrane catalyses a nH^+/Ca^{2+} exchange. *Plant Physiology*, **83**, 994-1000.

RASI-CALDOGNO, F., PUGLIARELLO, M.C., OLIVARI, C., and DE MICHELIS, M.I., 1989a. Identification and characterisation of the Ca^{2+}-ATPase which drives active transport of Ca^{2+} at the plasma membrane of radish seedlings. *Plant Physiology*, **90**, 1429-1434.

RASI-CALDOGNO, F., PUGLIARELLO, M.C., OLIVARI, C., and DE MICHELIS, M.I., 1989b. The calcium pump of the plasma membrane of higher plants. In *Plant Membrane Transport: the Current Position*. Eds J. Dainty, M.I. De Michelis, E. Marrè, and F. Rasi-Caldogno. Elsevier, Amsterdam. pp 225-230.

RASI-CALDOGNO, F., PUGLIARELLO, M.C., OLIVARI, C., and DE MICHELIS, M.I., GAMBARINI, G., COLOMBO, P., and TOSI, G., 1990. The plasma membrane Ca^{2+}-pump of plant cells: a radiation inactivation study. *Botanica Acta*, **103**, 39-41.

ROBINSON, C., LARSSON, C., and BUCKOUT, T.J., 1988. Identification of a calmodulin-stimulated $(Ca^{2+}-Mg^{2+})$-ATPase in a plasma membrane fraction isolated from maize (*Zea Mays*) leaves. *Physiologia Plantarum*, **72**, 177-184.

TESTER, M., 1990. Plant ion channels: whole-cell and single-channel studies. *New Phytologist*, **114**, 305-340.

WILLIAMS, L.E., SCHUELER, S.B., and BRISKIN, D.P., 1990. Further characterisation of the red beet plasma membrane Ca^{2+}-ATPase using GTP as an alternative substrate. *Plant Physiology*, **92**, 747-754.

ZOCCHI, G., 1988. Separation of membrane vesicles from maize roots having different calcium transport activities. *Plant Science Letters*, **54**, 103-107.

Secondary Ion and Metabolite Transporters

Introduction

While the primary active transport mechanisms are the engines which drive all nutrient accumulation and discharge in cells, the secondary transporters respond much more specifically to the nutritional status and needs of cells. They seem to be controlled in several ways, by expression, by allosteric regulation and by transmembrane pH gradients. It seems generally agreed that transporters do their work much faster than the more cumbersome ion pumps and it is expected that their abundance in the membrane will be correspondingly less. Thus, detection of the proteins themselves may be very difficult by conventional electrophoretic separation. This difficulty is compounded by the absence of convenient enzyme assays to verify the presence of the protein during separation and purification. The only recourse is to use functional transport assays in membrane vesicles - not always easy to set up. These are not excuses for lack of progress, for, as chapters in this section make clear, there have been most impressive advances in our knowledge of some of these proteins. Two factors operate to the advantage of researchers in this subject. First, many transporters appear to be strongly induced or repressed by the nutritional state of the organism (see Sauer) and second, mutations affecting the transporter may not be lethal. The latter opens up the opportunity for phenotypic complementation of mutants from genomic libraries (e.g. from *Arabidopsis*) using transformable species.

The deduced structures of transporters resemble in several major respects the ATPases and the PPase considered in the previous section. Hydrophobic, membrane-spanning α helices probably determine a pore in which there are specific binding domains for ions and metabolites. The process of probing the essential residues in these sites, by chemical modification and site-directed mutagenesis has only just begun. Kinetic models suggest orderly binding of H^+ and transported species to the transporter, but, elegant as they are, they do not tell us what actually happens during transport. Conformation changes in the polypeptide chain are usually assumed to bring about transport, but this is conjecture, and will remain so until the actual packing of the α helices is known. At the time of writing, despite the beauties of computer modelling (Wallmeier *et al.*), such knowledge is distant.

During discussion the nature of the induction and the general regulation of transporter activities were discussed. Where there is no constitutive level of activity it is clear that induction must involve gene transcription and protein synthesis. With sugar transporters this may be the way in which regulation occurs since, unlike their animal counterparts, they do not have any glycosylation sites nor do they undergo post-translational modification (Sauer). Where so few transporters have been identified it is not possible to comment on phosphorylation as a general method of regulation; in the chloroplast P_i translocator, however, phosphorylation does not seem to happen (Flügge). The putative Na^+/H^+ antiporter of the tonoplast (DuPont) is rapidly induced (maximally in 30 minutes) in conditions where cycloheximide blocks protein synthesis. This suggests some modification of a pre-existing tonoplast protein, perhaps *in situ*. Interest in this discussion arises because the relationships which exist between growth and nutrient intake, especially in higher plants, are not understood at a mechanistic level even though the relationships themselves have been apparent for many years.

Proton-Sugar Co-transporters in Plants

Norbert Sauer

Lehrstuhl für Zellbiologie und Pflanzenphysiologie
Universität Regensburg
Universitätsstrasse 31
D-8400 Regensburg, Germany

1.   Introduction

Sucrose and starch are the major end products of photosynthesis. Sucrose is actively loaded into the phloem and exported to those parts of the plant having a higher demand for carbohydrate (the sink tissues) whilst starch is accumulated in the chloroplasts of the mature leaves (the source tissues). Depending on the needs of the plant, the starch can be degraded to glucose-1-phosphate and finally leave the chloroplast as triose phosphate, which after conversion to sucrose in the cytoplasm, can also be loaded into the phloem by the sucrose carrier. There is evidence that the uptake of sugar into the cells of the sink tissues also occurs either via a sucrose carrier or, after degradation of sucrose to glucose and fructose, by cell wall invertases, via a monosaccharide carrier located in the plasmalemma of the respective cells. It is generally accepted that all of these plant plasma membrane sugar transporters are driven by the proton motive force.

Investigations of the transport properties of the plasma membranes of specific cell types or specific tissues of higher plants are difficult. The individual cells of most tissues have to be separated from each other by partial or complete degradation of the cell walls to make the transported substrates equally accessible to all of them. Another possibility is the isolation of pure plasma membranes (Larsson, Widell, and Kjellbom, 1987) and the determination of the transport properties of the resulting plasmalemma vesicles.

In our laboratory the unicellular green alga *Chlorella kessleri* has, for many years, proved to be a convenient model system for studying solute transport across the plasmalemma. It possesses an inducible transport system for hexoses, which can be turned on and off simply by the addition or removal of glucose to the growth medium (Tanner, 1989). In this paper, the identification of the responsible gene and the characterisation of its product is summarised. In addition, we show that this lower plant gene can serve as a valuable probe to screen higher plant libraries for genes and cDNAs, possibly involved in the different transport processes mentioned above. The sequences of two plant H^+/sugar co-transporters

Transport and Receptor Proteins of Plant Membranes
Edited by D.T. Cooke and D.T. Clarkson, Plenum Press, New York, 1992

67

are compared to transporters from other organisms. Differences and similarities are discussed.

2. Identification of Transport Proteins or their Genes

The inducibility of the transport system was found to be extremely helpful for the identification of both the putative H^+/glucose co-transporter protein and the respective cDNA from *Chlorella*. Detergent extracts from radiolabelled cells induced, or not, for glucose uptake showed clear differences in the protein pattern on SDS-polyacrylamide gels (Fenzl, Decker, Haass, and Tanner, 1977). Two bands could be identified in extracts from enriched plasma membranes, one at an apparent molecular weight of above 47,000 daltons, the second as a rather diffuse smear at about 40,000 to 42,000 daltons (Sauer and Tanner, 1984).

With this knowledge we started the differential screening of a cDNA library raised from polyA$^+$-RNA isolated from induced *Chlorella* cells (Sauer and Tanner, 1989). The screening gave a large number of cDNA clones coding for putatively glucose induced gene products. These isolated clones could be divided into 10 complementation groups. Sequencing and comparison with available data banks showed that two of these complementation groups encoded proteins with reasonable homology to the rat brain glucose carrier: pTF9 and pTF14 (Birnbaum, Haspel, and Rosen, 1986).

Screening for the respective genomic sequences in a *Chlorella kessleri* genomic library (Wolf, Tanner, and Sauer, 1991) yielded three different genes for putative transport proteins HUP1, HUP2, and HUP3 (HUP = hexose uptake). However, the third gene does not seem to be expressed after glucose treatment (Wolf and Sauer, unpublished results).

A similar situation was found when the HUP1 cDNA clone was used for the heterologous screening of cDNA and genomic libraries from *Arabidopsis thaliana* (Sauer, Friedländer, and Gräml-Wicke, 1990; Sauer, Friedländer, and Illig, unpublished results). Three different cDNAs and four different genes have been isolated up to now (STP1 - STP4; STP - sugar transport protein).

Using differential labelling of highly purified plasma membranes from sugar-beet leaves with radiolabelled N-ethylmaleimide (NEM) in the presence or absence of sucrose or its analogue palatinose Gallet, Lemoine, Larsson, and Delrot (1989) were able to detect a protein with an apparent molecular weight of 42,000 daltons. Antibodies raised against a fraction containing this protein inhibited sucrose transport into protoplasts. Transport of hexoses in the same system was not influenced by these antibodies (Lemoine, Delrot, Gallet, and Larsson, 1989). Sequences of the purified protein should provide specific probes to clone the respective cDNA or genomic clones.

3. Heterologous Expression

Cloning of cDNA or genomic clones with the help of heterologous DNA fragments as hybridisation probes, as with *Arabidopsis thaliana*, is usually the easiest method to use. However, in these cases a careful investigation of the function and properties of the product of the clones gene has to be performed for the following reasons: Firstly, a high degree of similarity alone does not necessarily predict anything about the function of a protein and, secondly, almost identical isozymes can be found in different compartments of a cell, or in different tissues of multicellular organisms.

The differential screening of the *Chlorella kessleri* cDNA library yielded two very similar cDNA clones for putative sugar transport proteins, pTF14, (Sauer and Tanner, 1989) and pTF9, (Sauer, unpublished data). mRNA levels for both clones increased upon addition of glucose to the medium. To determine which of the two clones encoded the monosaccharide transporter of the plasmalemma, we tried heterologous expression in several organisms. *Chlorella kessleri* itself has not yet been transformed successfully and purified protein for reconstitution experiments is not available.

Using the fission yeast *Schizosaccharomyces pombe* and the expression vector pEVP11 (Beach and Nurse, 1981; Russel and Nurse, 1986) we were able to show that the cDNA clone pTF14 and the corresponding gene (Wolf, Sauer, and Tanner, 1991) code for the *Chlorella* H^+/hexose co-transporter (Sauer, Caspari, Klebl, and Tanner, 1991) because:

- The additional transport activity in transformed *S. pombe* cells was found in the plasmalemma and polyclonal antibodies raised against the C-terminal end of the protein encoded by pTF14 bound to an antigen on the surface of transformed cells (Laub and Sauer, unpublished results);
- the substrate specificity of the transporter expressed in the fission yeast was identical to the substrate specificity of the carrier protein found in induced *Chlorella* cells (D-glucose, 3-O-methylglucose, D-fructose, D-galactose, D-xylose, D-arabinose);
- the K_m values of the transport protein expressed in *S. pombe* for various substrates were identical to the K_m values of the *Chlorella* H^+/hexose co-transporter for these substrates;
- uptake of the non-metabolisable substrate 3-O-methylglucose by transported *S. pombe* cells was clearly energy depended and driven by proton motive force;
- *in vitro* transcription/translation of a full length cDNA clone of pTF14 results in a ^{35}S-labelled protein running on SDS-polyacrylamide gels at an apparent molecular weight of 42,000 daltons (see above: Identification of transport proteins; Sauer, unpublished results) and transformed *S. pombe* cells expressed an additional protein with an apparent molecular weight of also 42,000 daltons (Laub and Sauer, unpublished results).

Schizosaccharomyces pombe proved to be an extremely convenient organism for two reasons:

a)   the K_m values of the endogenous sugar transporter for the various substrates tested are 500 to more than 1000-fold higher than the extremely low K_m values of the *Chlorella* carrier for the same substrates, and

b)   sugar transport across the plasma membrane of wild-type *S. pombe* cells occurs only as facilitated diffusion (non-metabolisable substrates cannot be accumulated inside the cells (Sauer *et al.*, 1990a)).

Thus, choosing the right (low) outside substrate concentration for uptake experiments allows us to determine the uptake rates of the transformed system with hardly any interference from the endogenous system. Most importantly, however, these experiments proved for the first time that both the uptake of sugar and protons are catalysed by one and the same transport protein which belongs to this group of homologous transporter.

The function and substrate specificities of the second *Chlorella* transporter (pTF9) could not be clarified to date.

The same expression system has been used for the investigation of a gene family encoding four different putative sugar transporters in *Arabidopsis thaliana* (Sauer *et al.*, 1990b; Sauer *et al.*, unpublished results). Full length cDNA clones are available for two members of this gene family. Both cDNAs were expressed in *S. pombe* using the same vector system. One of them (STP1) could be characterised as a proton/monosaccharide co-transporter with extreme functional similarity to the *Chlorella kessleri HUP1* gene. It is most strongly expressed in leaves, but the function of the STP1 protein in the *Arabidopsis* plant is not yet understood. The extreme sequence similarity of the two H^+/monosaccharide co-transporters, the fact that one of them (HUP1) is certainly located in the plasmalemma and the fact that the STP1 protein is found in the plasma membrane of transformed *S. pombe* cells makes it most likely that the STP1 protein is also located in the plasmalemma of the respective *Arabidopsis* cells. The final proof for this, however, has to be made using STP1-specific antibodies and *in situ* labelling of the protein in the plant.

4.   Sequence Similarities with Other Transporters

Sugar transporters were identified, cloned and sequenced in many different organisms from bacteria to man. Comparison of the different sequences revealed the existence of at least three groups of transport proteins with little or no homology between each other. The lactose permease of *E. coli* (Büchel, Gronnenborn, and Müller-Hill, 1980) and the intestinal Na^+/glucose co-transporters from rabbit and man (Hediger, Coady, Ikeda, and Wright, 1987; Hediger, Turk, and Wright, 1989) represent two of these different groups. The mammalian Na^+/glucose co-transporters were shown to be very similar to the *E. coli* Na^+/proline co-transporter (Nakao, Yamato, and Anraku, 1987) but not to any of the known sequences of sugar transporters. However, Rausch, Raszeja-Specht, and Koepsell (1989) found cross-reactivity of a plant H^+/glucose co-transporter with to the intestinal Na^+/glucose co-transporter, suggesting that higher plant sugar transporters might belong to this group of transport proteins. Twelve putative membrane spanning domains are assumed for the *E. coli* lactose permease with both the N- and the C-terminus on the cytoplasmic side (Kaback, 1989). The number of trans-membrane domains suggested for the intestinal transporters varies between eleven and twelve (Nakao *et al.*, 1987; Hediger *et al.*, 1989).

A third, much larger group of homologous sugar transport proteins is widely dispersed amongst different organisms and contains proteins like the facilitating glucose transporters from man (Mueckler, Caruso, Baldwin, Panico, Blench, Morris, Allard, Lienhard, and Lodish, 1985), the active H^+/xylose or arabinose co-transporters from *E. coli* (Maiden, Davis, Baldwin, Moore, and Henderson, 1987) and even transporters for disaccharides like the lactose transporter from *Klyveromyces lactis* (Chang and Dickson, 1988) or the maltose transporter from *Saccharomyces carlsbergensis* (Yao, Solliti, and Marmur, 1988).

In Table 1 the translated sequences of the *Arabidopsis thaliana (Arabi-Glc)* and the *Chlorella kessleri (Chlor-Glc)* H^+/monosaccharide co-transporters are compared with the sequences of the human hepatoma/erythrocyte glucose transporter *Hmery-Glc*, (Mueckler *et al.*, 1985), the *E. coli* H^+/arabinose co-transporter *Ecoli-Ara*, (Maiden *et al.*, 1987), the lactose transporter from *K. lactis*, *Klyve-Lac*, (Change and Dickson, 1988) and the maltose transporter from *S. carlsbergensis*, *Yeast-Mal*, (Yao *et al.*, 1988). Both plant sugar transporters exhibit clear sequence homologies to the other transporters of this group (the most conserved amino acids being marked with an asterisk), which all seem to possess 12 membrane spanning domains (M1-M12). It is assumed that both ends of the mature protein are located on the cytoplasmic side of the plasma membrane, as is the case in the *E. coli* lactose permease. The highly conserved sequence which is found between the putative trans-membrane helices M2 and M3 and the helices M8 and M9 is:

RXGRR

with R being sometimes replaced by lysine and X standing for an amino acid with a hydrophobic (frequently aromatic) side chain. This sequence is likely to form a β-turn between the two trans-membrane helices and to interact with negative charges of the polar head groups of membrane lipids. Interestingly this sequence is also found at the same positions in the otherwise unrelated *E. coli* lactose permease.

A second set of highly conserved sequences is found after the membrane spanning helices M6 and M12. The respective motifs

PESP and PETKG

are almost unchanged in most of the transporters of this third group. However, the function of these motifs is unclear and remains to be investigated by site directed mutagenesis.

Besides the general conservation of amino acids with lipophilic side chains throughout most of the putative transmembrane helices there is a conserved motif in the putative transmembrane helix M7:

QQ---GIN---YY

with the first Y replaced by F in the plant sugar carriers. This strong conservation of asparagine and glutamine residues within a putative membrane spanning region could perhaps be explained by an involvement of this helix in the formation of a hydrophilic pore through the membrane. Photoaffinity labelling of the human erythrocyte glucose transporter with radio-labelled Cytochalasin B, an inhibitor of mammalian sugar transporters (Carter-Su, Pessin, Mara, Gitamer, and Czech, 1982), and the combination of this method with the comparison of proteolytic fragment patterns (Holman and Rees, 1987), suggested that the trans-membrane helices M7 - M10 might form this hydrophilic pore.

In addition to these homologies which are common to all transporters of this rather related group, there is a highly conserved area which is only present in the different plant sugar transporters and in the members of their respective gene families in *Chlorella* and *Arabidopsis* (Sauer, unpublished data). This sequence is found after helix M12 and contains the PETKG motif, which is much longer here:

PETKG - PIE -------- HWYW - R

with the Y sometimes replaced by F. As mentioned before, the function of this conserved region is not clear at present.

As well as the three different groups of sugar transporters and sugar/ion co-transporters described above, additional bacterial sugar transporters seem to exist which cannot be assigned to any of these three groups (reviewed by Henderson, 1990).

5.   Summary

This work summarises the current knowledge of the sequences and structure of plant H^+/sugar co-transport proteins. There seem to be similar families of putative transport proteins in lower and higher plants. All of the sequences of plant transport proteins obtained so far show that these proteins belong to a large group of homologous transporters found in bacteria, yeast and mammals. The individual members of this group are rather variable regarding their energy dependence, substrate specificity and substrate size. At present, it is unclear whether other H$^+$/sugar co-transporters, which are homologous to the intestinal Na$^+$/glucose co-transporters, exist in higher plants as suggested by data from Rausch *et al.*, (1989). However, the fact that the intestinal Na$^+$/glucose co-transporters are related to a bacterial Na$^+$/proline co-transporter (Nakao *et al.*, 1987) could mean that the specificity for Na$^+$-ions is the primary reason for this structural and sequence homology. In this case one would not expect to find similar proteins in plants.

The finding that transporters for disaccharides like maltose and lactose have been identified as members of this homologous family of transporters is also very promising. The disaccharide, sucrose, is one of the most important sugars in the long distance transport of higher plants and there are still many open questions about the mechanism(s) of phloem loading and unloading. One can hope that a putative H$^+$/sucrose co-transporter is either amongst the clones that have already been obtained and not yet characterised from *Arabidopsis thaliana*, or that the available clones might at least serve as probes in the search for this protein.

The identification, characterisation and localisation of the different sugar transporters in a single plant is essential to the understanding of the distribution of carbon amongst the different tissues and will help to elucidate the often obscure ways solutes have to go from one plant cell to the other.

Table 1. Comparison of the translated sequences of the H$^+$/monosaccharide co-transporters from *Chlorella kessleri (HUP1)* and *Arabidopsis thaliana (STP1)*, the glucose transporter from man *(HepG2)*, the H$^+$/arabinose co-transporter from *E. coli (AraE)*, the lactose transporter from *Klyveromyces lactis (LAC12)* and the maltose transporter from *Saccharomyces carlsbergensis (MAL6T)*. Conserved parts of the sequences are marked with asterisks; the putative transmembrane helices of the *Chlorella* carrier are marked (M1 - M12).

```
Conserved   ..........  ..........  ..........  ..........  ..........
Chlor-Glc   ..........  ..........  ........MA  GGGVVVVSGR  G........L   14
Arabi-Glc   ..........  ..........  ........MP  AGGFVVGDGO  ....K.....   13
Hmery-Glc   ..........  ..........  ..........  ..........  ..MEPSSKKL    8
Ecoli-Ara   ..........  ..........  ..........  ..........  .........M    1
Klyve-Lac   MADHSSSSSS  LQKKPINTIE  HKDTLGNDRD  HKEALNSDND  NTSGLKINGV   50
Yeast-Mal   ..........  ..........  ........MK  GLSSLINRKK  DRNDSHLDEI   22

                                    M1
Conserved   .........  ..........  ..........  ......***.  **........
Chlor-Glc   STGDYRGGLT  VYVVM.....  ....VAFMAA  CGGLLLGYDN  GVTGGVVSLE   55
Arabi-Glc   ...AYPGKLT  PFVLF.....  ....TCVVAA  MGGLIFGYDI  GISGGVTSMP   51
Hmery-Glc   .TGRLMLAVG  GAVL......  ..........  .GSLQFGYNT  GVINAR...Q   37
Ecoli-Ara   VTINTESALT  PRSLRDTRRM  NM.FVSVAAA  VAGLLFGLDI  GVIAGALP..   48
Klyve-Lac   PIEDAREEVL  LPGYLSKQYY  KLYGLCFITY  LCATMQGYDG  ALMGSIYTED  100
Yeast-Mal   ENGVNATEFN  SIEMEEQGKK  SDFDLSHLEY  GPGSLIPNDN  NEEVPDLLDE   72

Conserved   .....**...  ..........  ..........  .......... . ..........
Chlor-Glc   AFEKKFFPDV  WAKKQEVHED  SPYCTYDNAK  LQLFVSSLFL  AGLVSCLFAS  105
Arabi-Glc   SFLKRFFPSV  YRKQQEDAST  NQYCQYDSPT  LTMFTSSLYL  AALISSLVAS  101
Hmery-Glc   KVIEEFYNQT  WVHR...YGE  SILPTTLTTL  WSLSVAIFSV  GGMIGSFSVG   84
Ecoli-Ara   .....FITDH  FVLTSRLQE.  ..........  ..WVSSMML  GAAIGALFNG   80
Klyve-Lac   AYL.KYYHLD  INSSSGT...  ..........  .GLVFSIFNV  GQICGAFFVP  135
Yeast-Mal   AMQDAKEADE  SERGMPLMTA  LKTYPKAAAW  SLLVSTTLIQ  EGYDTAILGA  122

            M2                                              M3
Conserved   ..........  .........*  ****......  ....**.***  ..........
Chlor-Glc   WI........  .......TRN  WGRKVTMGIG  GAF.FVAGGL  VN..AFAQDM  137
Arabi-Glc   TV........  .......TRK  FGRRLSMLFG  GIL.FCAGAL  IN..GFAKHV  133
Hmery-Glc   LF........  .......VNR  FGRRNSMLMM  NLLAFVSAVL  MGFSKLGKSF  119
Ecoli-Ara   WL........  .......SFR  LGRKYSL.MA  GAILFVLGSI  GS..AFATSV  112
Klyve-Lac   LM........  .......DWK  .GRKPAILI.  GCLGVVIGAI  ISSLTTTK..  166
Yeast-Mal   FYALPVFQKK  YGSLNSNTGD  YEISVSWQIG  LCLCYMAGEI  VGLQVTGPSV  172
```

```
                                         M4
Conserved   .......***  ******...  ....:....:  ...**.....  ..........
Chlor-Glc   ...A...MLI  VGRVLLG...  ....FGVGLG  SQVVPQ....  ..........  160
Arabi-Glc   .W.....MLI  VGRILLG...  ....FGIGFA  NQAVP.....  ..........  155
Hmery-Glc   ...E...MLI  LGRFIIGVY.  ......CGLT  TGFVP.....  ..........  141
Ecoli-Ara   ...E...MLI  AARVVLG...  ....IAVGIA  SYTAP.....  ..........  134
Klyve-Lac   ...S...ALI  GGRWFVA...  ....FFATIA  NAAAP.....  ..........  188
Yeast-Mal   DYMGNRYTLI  MALFFLAAFI  FILYFCKSLG  MIAVGQALCG  MPWGCFQCLT   22


                                         M5
Conserved   ...*****.   .***.*....  **....****  *.........  ........**
Chlor-Glc   ..YLSEVAPF  SHRGMLNIGY  QLFVTIGILI  AGL.VN.Y..  AVRDWENGWR  204
Arabi-Glc   .LYLSEMAPY  KYRGALNIGF  QLSITIGILV  AEV.LNYFFA  KIKGGWG.WR  202
Hmery-Glc   .MYVGEVSPT  AFRGALGTLH  QLGIVVGILI  AQV.FG.L.D  SIMGNKDLWP  187
Ecoli-Ara   .LYLSEMASE  NVRGKMISMY  QLMVTLGIVL  AFLSD....T  AF.SYSGNWR  178
Klyve-Lac   .TYCAEVAPA  HLRGKVAGLY  NTLWSVGSIV  AAFSTYGTNK  NFPNSSKAFK  178
Yeast-Mal   VSYASEICPL  ALRYYLTTYS  NLCWTFGQLF  AAGIMKNSQN  KYANSELGYK  272


            M6
Conserved   *.*.....**  *.*.*...**  ....*****.  ....:**...  .........*.
Chlor-Glc   LSLGLAA.AP  GAILFLGSLV  L...PESPN.  ...FLVEKGK  TEKGREVLQK  246
Arabi-Glc   LSLGGAV.VP  ALIITIGSLV  L...PDTPNS  ...MIERGO   HEEAKTKLRR  244
Hmery-Glc   LLLSIIF.IP  A..L.LWCIV  LPFCPESPRF  LLINRNEENR  AKSVLKKLRG  233
Ecoli-Ara   AMLGVLA.LP  AVLLIILVVF  L...PNSPR.  ...WLAEKGR  HIEAEEVLRM  220
Klyve-Lac   IPLYLQMMFP  GLVCIFGWLI  ....PESPR.  WLVGVR...E  EEAREFIIK.  279
Yeast-Mal   LPFALQWIWP  LPLAVGIFLA  ....PESP W  WLVKKER..I  DQARRSLERI  315


Conserved   ..........  ....*..*..  ..........  ..........  ....*...**
Chlor-Glc   LRGTSE....  .VDAEFADIV  A...AVEIAR  PITMRQSWAS  L.FTRR..YM  285
Arabi-Glc   IRGVDD....  .VSQEFDDLV  A...ASKESQ  SIE..HPWRN  L.LRRK..YR  281
Hmery-Glc   TADV...TH.  .DLQEMKEES  RQMMREKKVT  I......LEL  FRSPA...YR  272
Ecoli-Ara   LRDTSE....  KAREELNEI.  ...RESLKL..  ...KQGGWAL  FKINRN..VR  256
Klyve-Lac   YHLNGDRTHP  LLDMEMAEII  ESFHGTDLSN  PLEMLDVR.S  LFRTRSDRYR  328
Yeast-Mal   LSGKGPEKEL  LVSMELDKIK  T.....TIEK  EQKMSDEGTY  WDCVKDGINR  360


            M7
Conserved   ..........  **********  ..:**.....*.  ..........  .....*....
Chlor-Glc   PQLLSTFVIQ  FFQQFTGINA  IIFYVPVLFS  SLG.SANSAA  LLNTVVVGAV  334
Arabi-Glc   PHLTMAVMIP  FFQQLTGINV  IMFYAPVLFN  TIGFTTD.AS  LMSAVVTGSV  330
Hmery-Glc   QPILIAVVLQ  LSQQLSGINA  VFYYSTSIFE  KAGVQQPVYA  TIGSGIVNTA  319
Ecoli-Ara   RAVFLGMLLQ  AMQQFTGMNI  IMYYAPRIFK  MAGFTTTEQQ  MIATLVVGLT  306
Klyve-Lac   AMLV..ILMA  WFGQFSGNNV  CSYYLPTMLR  NVGMKSVSLN  VLMNGVYSIV  376
Yeast-Mal   RRTRIACLC.  WIGQCSCGAS  LIGYSTYFYE  KAGVSTD.TA  FTFSIIQYCL  408


            M8                                M9
Conserved   ....*..*.*  .***.***..  .......*...  ...:...***
Chlor-Glc   NVGSTLIAVM  FSDKF.GRRF  LLIEGGI..Q  CCLAMLTTG.  VVLAIEFAKY  380
Arabi-Glc   NVGATLVSIY  GVDRW.GRRF  LFLEGGT..Q  ....MLICQA  VVAACIGAKF  373
Hmery-Glc   ...FTVVSLF  VVERA.GRRT  LHLI.GLA..  .GMAGCAILM  TIALALLEQL  361
Ecoli-Ara   FMFATFIAVF  TVDKA.GRK.  PALKIGFS..  .VMALGTLVL  ...GYCLMQF  348
Klyve-Lac   TWISSICGAF  FIDKI.GRRE  GFL..GSISG  AALAL.....  TGLSICTARY  418
Yeast-Mal   GIAATFVS.W  WASKYCGRFD  LYA.FGLAFQ  AIMFFIIGGL  GCSDTHGAKM  456


            M10
Conserved   ..........  .:....:...  ...*......  ***.....*.
Chlor-Glc   GTDP....LP  KAVASGILAV  ICIFISGFAW  SWGPMGWLIP  SEIFTLETRP  426
Arabi-Glc   GVDGTPGELP  KWYAIVVVTF  ICIYVAGFAW  SWGPLGWLVP  SEIFPLEIRS  423
Hmery-Glc   PWMSYL....  ..SIVAIFG.  FVAFFEVGPG  PI.PW.FIV.  AELFSQGPRP  401
Ecoli-Ara   DNGTASSGLS  W..LSVGMTM  MC..IAGYAM  SAAPVVWILC  SEIQPLKCRD  394
Klyve-Lac   EKTK.....K  KSASNGALVF  IYLFGGIFSF  AFTPMQSMYS  TEVSTNLTRS  463
Yeast-Mal   GSGALLMVVA  FFYNLGIAPV  VFCLVSEMPS  SR......LR  TKTIIL.ARN  499
```

```
                        M11                              M12
Conserved   ........... ....**.... *.......... ........... ...........
Chlor-Glc   AGTAVAVVGN  FLFSFVIGQA  F..VSMLCAM  EYGVFLFFAG  WLVIMVLCAI  474
Arabi-Glc   AAQSITVSVN  MIFTFIIAQI  F..LTMLCHL  KFGLFLVSAF  FVVVMSIFVY  471
Hmery-Glc   AAIAVAGFSN  WTSNFIVGMC  FQYVEQLCGP  Y..VFIIFTV  LLVLFFIFTY  449
Ecoli-Ara   FGITCSTTTN  WVSNMIIGAT  FLTL.LDSIG  AAGTFWLYTA  LNIAFVGITF  443
Klyve-Lac   KAQLLNFVVS  GVAQFV..NQ  FATPKAMKNI  KYWFYVFYVF  FDIFEFIVIY  511
Yeast-Mal   ANYVIQVVVT  VLIMYQLNSE  KWNWGAKSG.  .....FFWGG  FCLCATLAWA  543


Conserved   ...****** . ......***. .......... .......... ..........
Chlor-Glc   FL.LPETKGV  PI....ERVQ  .ALYARHWFW  NRV....MGP  AAAEVIAEDE  514
Arabi-Glc   IF.LPETKGI  PI....EEMG  .QVWRSHWYW  SRFVED..GE  YGNALEMGKN  513
Hmery-Glc   FK.VPETKGR  TF....DEIA  SGF.......  RQGGASQSDK  TPEELF....  483
Ecoli-Ara   WL.IPETKNV  TL....EHIE  RKLMAGEKLR  NIGV*.....  ..........  472
Klyve-Lac   FFFV.ETKGR  SR....EELE  VVFEAPNPRK  ASVDQAFLAQ  VRATLVQRND  556
Yeast-Mal   VVDLPETAGR  TFIEINELFR  LGVPARKFKS  TKVDPFAAAK  AAAAEINVKD  593


Conserved   .......... .......... .......... ...
Chlor-Glc   KRVAAASAII  KEEELSKAMK  *.........  ...                      534
Arabi-Glc   SNQAGTKHV*  ..........  ..........  ...                      522
Hmery-Glc   HPLGADSQV*  ..........  ..........  ...                      492
Ecoli-Ara   ..........  ..........  ..........  ...                      472
Klyve-Lac   VRVANAQN.L  KEQEPLKSDA  DHVEKLSEAE  SV*                      587
Yeast-Mal   PKEDLETSVV  .DEGRSTPSV  VNK*......  ...                      615
```

Acknowledgement

I want to thank Prof. Dr. W. Tanner for his help and continuous discussions. This work was supported by the Deutsche Forschungsgemeinschaft (SFB43-C5).

REFERENCES

BEACH, D. and NURSE, P., 1981. High-frequency transformation of the fission yeast *Schizosaccharomyces pombe*. *Nature*, **290**, 140-142.

BIRNBAUM, M.J., HASPEL, H.C., and ROSEN, O.H., 1986. Cloning and characterisation of a cDNA encoding the rat brain glucose-transporter protein. *Proceedings of the National Academy of Sciences, USA*, **83**, 5784-5788.

BÜCHEL, D.E., GRONENBORN, B., and MÜLLER-HILL, B., 1980. Sequence of the lactose permease gene. *Nature*, **283**, 541-545.

CARTER-SU, C., PESSIN, J.E., MARA, R., GITAMER, W., and CZECH, M.P., 1982. Photoaffinity labelling of the human erythrocyte glucose transporter. *Journal of Biological Chemistry*, **257**, 5419-5425.

CHANG, U.-D. and DICKSON, R.C., 1988. Primary structure of the lactose permease gene from the yeast *Klyveromyces lactis*. Presence of an unusual transcript structure. *Journal of Biological Chemistry*, **263**, 16696-16703.

FENZL, F., DECKER, M., HAASS, D., and TANNER, W., 1977. Characterisation and partial purification of an inducible protein related to hexose proton co-transport of *Chlorella vulgaris*. *European Journal of Biochemistry*, **72**, 509-514.

GALLET, O., LEMOINE, R., LARSSON, C., and DELROT, S., 1989. The sucrose carrier of the plant plasma membrane. I. Differential affinity labelling. *Biochimica et Biophysica Acta*, **978**, 56-64.

HEDIGER, M.A., COADY, M.J., IKEDA, T.S., and WRIGHT, E.M., 1987. Expression cloning and cDNA sequencing of the Na⁺/glucose co-transporter. *Nature*, **330**, 379-381.

HEDIGER, M.A., TURK, E., and WRIGHT, E.M., 1989. Homology of the human intestinal Na⁺ glucose and *Escherichia coli* Na⁺/proline co-transporter. *Proceedings of the National Academy of Sciences, USA*, **86**, 5748-5752.

HENDERSON, P.J.F., 1990. Proton-linked sugar transport systems in Bacteria. *Journal of Biomembranes and Bioenergetics*, **22**, 526-569.

HOLMAN, G.D., and REES, W.D., 1987. Photolabelling of the hexose transporter of external and internal sites: fragmentation patterns and evidence for a conformational change. *Biochimica et Biophysica Acta*, **897**, 395-405.

KABACK, H.R., 1989. Molecular biology of active transport: from membrane to molecule to mechanism. *Harvey Lectures*, **83**, 77-105.

LARSSON, C., WIDELL, S., and KJELLBOM, P., 1987. Preparation of high purity plasma membranes. *Methods in Enzymology*, **148**, 558-568.

LEMOINE, R., DELROT, S., GALLET, O., and LARSSON, C., 1989. The sucrose carrier of the plant plasma membrane. II. Immunological characterisation. *Biochimica et Biophysica Acta*, **978**, 65-71.

MAIDEN, M.C.J., DAVIS, E.O., BALDWIN, S.A., MOORE, D.C.M., and HENDERSON, P.J.F., 1987. Mammalian and bacterial sugar transport proteins are homologous. *Nature*, **325**, 641-643.

MUECKLER, M., CARUSO, C., BALDWIN, S.A., PANICO, M., BLENCH, I., MORRIS, H.R., ALLARD, W.J., LEINHARD, G.E., and LODISH, H., 1985. Sequence and structure of a human glucose transporter. *Science*, **229**, 941-945.

NAKAO, T., YAMOTO, I., and ANRAKU, Y., 1987. Nucleic acid sequence of *putP*, the proline carrier gene of *Escherichia coli* K12. *Molecular and General Genetics*, **208**, 70-75.

RAUSCH, T., RASZEJA-SPECHT, A., and KOEPSELL, H., 1989. Identification of an M_r 75,000 component of the H^+/D-glucose co-transporter from *Zea mays* with monoclonal antibodies directed against the mammalian Na^+/D-glucose co-transporter. *Biochimica et Biophysica Acta*, **985**, 133-138.

RUSSEL, P., and NURSE, P., 1986. *cdc25$^+$* functions as an inducer in the mitotic control of fission yeast. *Cell*, **45**, 145-153.

SAUER, N., CASPARI, T., KLEBL, F., and TANNER, W., 1990a. Functional expression of the *Chlorella* hexose transporter in *Schizosaccharomyces pombe*. *Proceedings of the National Academy of Sciences, USA*. **87**, 7949-7952.

SAUER, N., FRIEDLÄNDER, K., and GRÄML-WICKE, U., 1990b. Primary structure, genomic organisation and heterologous expression of glucose transporter from *Arabidopsis thaliana*. *The EMBO Journal*, **9**, 3045-3050.

SAUER, N., and TANNER, W., 1984. Partial purification and characterisation of inducible transport proteins of *Chlorella*. *Zietschrift für Pflanzenphysiologie*, **114**, 367-375.

SAUER, N., and TANNER, W., 1989. The hexose carrier from *Chlorella*. cDNA cloning of a eucaryotic H^+-co-transporter. *FEBS Letters*, **259**, 43-46.

TANNER, W., 1969. Light driven active uptake of 3-O-methylglucose via an inducible hexose uptake system of *Chlorella*. *Biochemical and Biophysical Research Communications*, **36**, 278-283.

WOLF, K., SAUER, N., and TANNER, W., 1991. The *Chlorella* H^+/hexose co-transporter gene. *Current genetics*, **19**, 215-219.

YAO, B., SOLLITTI, P., and MARMUR, J., 1989. Primary structure of the maltose-permease-encoding gene of *Saccharomyces carlsbergensis*. *Gene*, **79**, 189-197.

Insights into the Structure of the Chloroplast Phosphate Translocator Protein

Holger Wallmeier[2], Andreas Weber[1], Armin Gross[1]
and Ulf-Ingo Flügge[1,3]

1.   Julius-von-Sachs-Institut für Biowissenschaften mit Botanischem Garten der
      Universität Würzburg
      Mittlerer Dallenbergweg 64
      8700 Würzburg, Germany

2.   Zentralforschung Hoechst AG
      Postfach 800 320
      6230 Frankfurt 80, Germany

3.   Author to whom correspondence should be addressed

1.   Physiological Function and Properties of the Phosphate Translocator from Plastids

The triose phosphate-3-phosphoglycerate-phosphate translocator (in short, phosphate translocator) can be regarded as the main transport function of the inner envelope membrane of C3-chloroplasts (for a recent review see Flügge and Heldt, 1991). It mediates a strict counter-exchange of the above mentioned metabolites thereby exporting the fixed carbon in the form of triose phosphate and 3-phosphoglycerate from the chloroplasts to the cytosol. Triose phosphate is then used for the synthesis of sucrose and 3-phosphoglycerate for that of amino acids. Sucrose and amino acids are the main products translocated via the sieve tubes to other parts of the plants. In C3-type chloroplasts the phosphate translocator accepts at its substrate binding site only phosphate and C3 compounds with the phosphate attached to the C3 atom.   However, that from C4-mesophyll chloroplasts also transports C3 compounds with the phosphate attached to the C2 atom i.e. phosphoenolpyruvate which is formed inside the chloroplasts and transferred via the C4-phosphate translocator to the cytosol to be carboxylated to oxaloacetate (Gross, Brückner Heldt, and Flügge, 1990). In amyloplasts, the phosphate translocator accepts at its binding site all the mentioned substrates and, in addition, glucose 6-phosphate as the substrate for starch biosynthesis and for the oxidative pentose phosphate pathway (Borchert,

Transport and Receptor Proteins of Plant Membranes
Edited by D.T. Cooke and D.T. Clarkson, Plenum Press, New York, 1992

77

Große, and Heldt, 1989). It becomes obvious that in different plastids the phosphate translocators of the inner envelope membrane can be rather different with respect to their transport specificities.

The translocator proteins from C3 plants has been studied most extensively. It had been identified as the major component of the inner envelope membrane with an apparent relative molecular mass of 29,000 (E29, spinach) and 30,000 (E30, pea) comprising about 10-15% of the total envelope membrane protein (Figure 1). Both proteins can specifically be labelled by treatment of the chloroplasts with micromolar amounts of tritiated dihydro-4,4'-diisothiocyanostilbene-2,2'-disulfonic acid ([^{3}H]H$_2$DIDS) (Figure 1) which effectively inhibits phosphate transport activity (Rumpho, Edwards, Yousif, and Keegstra, 1988; Gross et al., 1990).

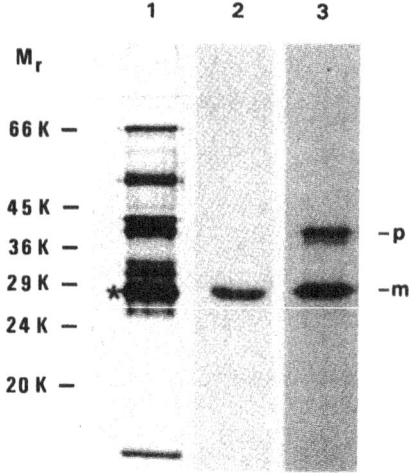

Figure 1 E29 which is encoded by the corresponding cDNA represents the major envelope membrane polypeptide and can be labelled by [^{3}H]DIDS. Chloroplasts were incubated with 2 μM [^{3}H]DIDS and the envelope membranes were analyzed by SDS-PAGE. Lane 1, silver-stained envelope membrane proteins; E29 is marked by an asterisk. Lane 2, fluorographic analysis of envelope membrane proteins. Lane 3, the cDNA clone encoding E29 was transcribed in vitro (Willey et al., 1990) and translated in the presence of [^{35}S]methionine, processed in vitro (Willey et al., 1990) and analysed by SDS-PAGE and fluorography. p, precursor protein; m, mature protein.

Hydrodynamic studies of the isolated protein and rotational diffusion measurements of the eosin-5-isothiocyanate-labelled reconstituted translocator protein led to the conclusion that the protein exists in its functional state as a dimer made up of two identical subunits. It is deeply embedded into the membrane and a value of about 1.7 nm was calculated for the radius of the protein in the membrane (Flügge, 1985; Wagner, Apley, Gross, and Flügge, 1989).

The phosphate translocator protein is encoded by nuclear genes and synthesised in the cytosol as a higher-molecular-weight precursor protein with a N-terminal extension, the transit peptide (Flügge and Wessel, 1984); it is subsequently inserted into the chloroplast envelope membrane and processed to its mature size (Flügge, Fischer, Gross, Sebald, Lottspeich, and Eckerskorn, 1989) which exhibits the same electrophoretic mobility as the authentic translocator protein (E29, Figure 1).

The entire cDNA sequence of the phosphate translocator precursor protein from both spinach and pea chloroplasts has recently been elucidated (Flügge et al., 1989; Willey, Fischer, Wachter, Link, and Flügge, 1990). Determination of the N-terminal sequence of the mature phosphate translocator proteins from both spinach and pea chloroplasts revealed that the transit peptides consist of 80- and 72-amino-acid residues, respectively. In contrast to the transit of precursor proteins that are destined to other chloroplastic sub-compartments, the transit peptide of the translocator contains a positively charged amphiphilic $\propto$-helix with a high hydrophobic moment (Willey et al., 1990). Possibly, such helical structures are sufficient to direct an inner envelope membrane protein to its membrane (Dreses-Werringloer, Fischer, Wachter, Link, and Flügge, 1990).

Unexpectedly, Blobel and co-workers suggested that E29/E30 represents the chloroplast protein import receptor and not the phosphate translocator (Pain, Kanwar, and Blobel, 1988; Schnell, Blobel, and Pain, 1990). However, we have presented evidence that strongly argues against this suggestion and which clearly identifies E29/E30 as the chloroplast phosphate translocator (Flügge, Weber, Fischer, Lottspeich, Eckerskorn, Waegemann, and Soll, 1991).

2. Molecular Organization of the Chloroplast Phosphate Translocator in the Membrane

The mature part of the spinach translocator protein contains 324 amino acid residues corresponding to a molecular mass of M_r 35,603 which is substantially higher than that determined by SDS-PAGE (29 kDa). This discrepancy might be explained by an unusually high binding of detergent resulting from the hydrophobic nature of the protein (Flügge, 1985). Indeed, the translocator contains a large excess of non-polar amino acid residues resulting in an overall polarity index of only 33%. The hydropathy plot for the amino acid sequence revealed the presence of 6-7 membrane-spanning stretches of 20-23 amino acid residues that can all form $\propto$-helical structures (Willey et al., 1990). These data leads to the tentative model of the arrangement of the monomeric phosphate translocator as depicted in Figure 2. An even number of membrane-spanning helical sections is suggested since both the N- and the C-terminus appear to be located on the inter-membrane (cytosolic) side of the protein (unpublished results). The trans-membrane helices are connected by hydrophilic loops of which all but two are rather long. It may be noted that the connecting polypeptide between helix 2 and helix 3 (Pro_{74} to Ser_{135}) may contain an additional $\propto$-helix (Val_{90} to Phe_{112}) that we suggest to be only membrane-associated.

The question now arises how these 12 transmembrane helices might be arranged in the dimeric translocator protein. A number of restrictions concerning the relative positioning of the helices can be postulated: (i), The loop between helix 1 and 2 consists of only 12 amino acid residues (Asn_{40} to Tyr_{51}) suggesting that these helices may always have a close association with one another, (ii), Helix 5 and 6 are connected by a very short loop of only three amino acids, their close association is therefore beyond any doubt; (iii), Both helix 2 and 4 contain a cysteine residue at about the same height (Cys_{65} and Cys_{207}, respectively). We assume it to be very likely that both helices are connected by a disulfide bridge although experimental evidence for this is still lacking.

Model structures of all six helices were then generated by using standard geometrical parameters for $\propto$-helices (Schultz and Schirmer, 1983). The results are shown in Figure 3. The distribution of the side chains on the surface of the individual helices was then analyzed and is represented schematically in Figure 4. It becomes obvious from this figure that most of the helices show a pronounced alternation between hydrophilic and hydrophobic regions. It is assumed that the hydrophilic side chains are directed towards the inside of the protein thus forming a hydrophilic translocation pore through which the substrates could be transported across the membrane. In the hydrophobic

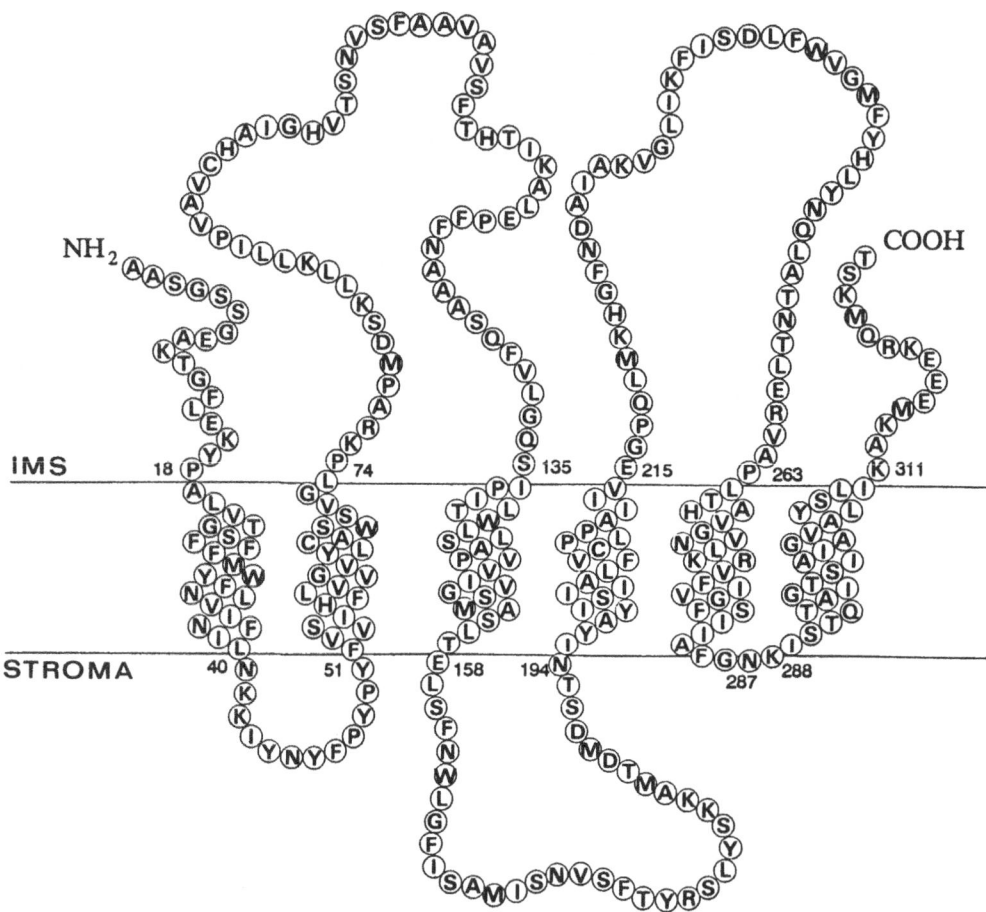

Figure 2 Model for the arrangement of the polypeptide chain of the monomeric chloroplast phosphate translocator within the lipid bilayer, based on the hydropathy plot (Kyte and Doolittle, 1982) and additional information. IMS, intermembrane space.

regions, areas with both aliphatic and aromatic side chains can be detected. The aliphatic side chains may preferentially interact with the hydrophobic (aliphatic) core of the lipid bilayer whereas the aromatic side chains may contribute to the stability of the protein structure itself via aromatic interactions between the individual helices. This appears feasible since for steric and entropic reasons a contact of similar side chains is more likely than a "mixed" contact between aromatic and aliphatic side chains. However, the individual helices in the dimeric protein are mainly held together via hydrogen bonds and interactions between aliphatic side chains. It is also obvious from Figures 3 and 4 that due to the presence of small-sized amino acid residues (Glycine and Proline) there are some "gaps" in the side chain environment of the helices. These "gaps" should serve to adopt bulky (aromatic) side chains of neighbouring helices thus stabilizing the whole structure by apolar interactions.

The above considerations suggest a schematic alignment of the 12 α-helices in the dimeric translocator protein as depicted in Figure 5. According to this picture the dimeric translocator exhibits an approximately C_2-symmetry with a C_2-rotational symmetry axis perpendicular to the membrane plane.

A three-dimensional structure according to this scheme was built by means of computer-graphics and was subsequently energy-minimized (see Figures 6-8) using the AMBER force field (Weiner, Kollman, Case, Singh, Ghio, Alagona, Profeta, and Weiner, 1984). In addition, a molecular dynamics simulation at 150K *in vacuo* was performed for further relaxation and to check for conformational inconsistencies. The system turned out to be stable under the given conditions.

In the plane of the membrane, the resulting structure has a total length of 5.0 nm and a width of 2.6 nm (Figure 7). These values almost fit those obtained by hydrodynamic and rotational diffusion studies of the isolated translocator protein (Flügge, 1985; Wagner *et al.*, 1989). The corresponding values for the hydrophilic translocation channel are 2.2 nm and 0.8 nm, respectively. It may be emphasised that the translocation channel is not just a hole within the protein but is buried by several amino acid residues directed towards the pore thus preventing unspecific permeation of substrates through the pore. Even though the missing loops definitely further stabilise the structure, there is a considerable flexibility in the alignment of the helices that may be important for physiological function of the pore. Consistent with this is the observation that some changes in the shape of the translocator occurred during transport function (Wagner *et al.*, 1989). Therefore, the above values can be taken only as approximations.

Earlier experiments had shown that an arginine and a lysine residue are involved in the binding of the twice-negatively charged substrates to the translocator (Flügge and Heldt, 1979). This lysine residue presumably represents the target for the inhibitors of the translocator e.g. pyridoxal 5'-phosphate and DIDS. Interestingly, Lys_{273} and Arg_{274} (helix 5) are the only charged amino acid residues contained in that part of the protein embedded within the membrane. It is suggested that these two cationic residues are involved in binding of the negatively charged substrates.

According to the three-dimensional structure of the translocator as presented here, the positively charged side chains of these two amino acid residues point to the center of the translocation channel (Figure 7). The distance between the centres of the guanidinium groups is roughly 1.0 nm and between the ϵ-amino groups of the lysine side chains 2.0 nm. Thus, an agent like DIDS should be able to interact with the two lysine residues in the dimer and to completely block the translocation channel (Figure 8).

As a consequence of the disulfide bridge between helices 2 and 4, helix 2 cannot provide polar side chains for the surface of the translocation pore. The resulting hydrophobic area inside the pore obviously assists to orientate the side chains of Phe_{25} and Trp_{29} (helix 1) and Tyr_{64} (helix 2). These three side chains are found in a stack-like arrangement so that they cannot move independently (see Figure 7A).

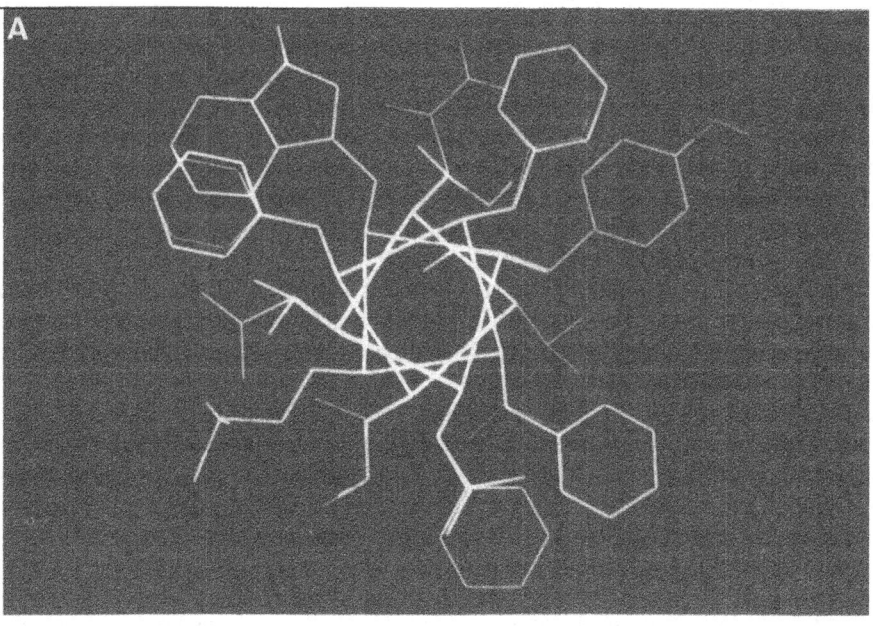

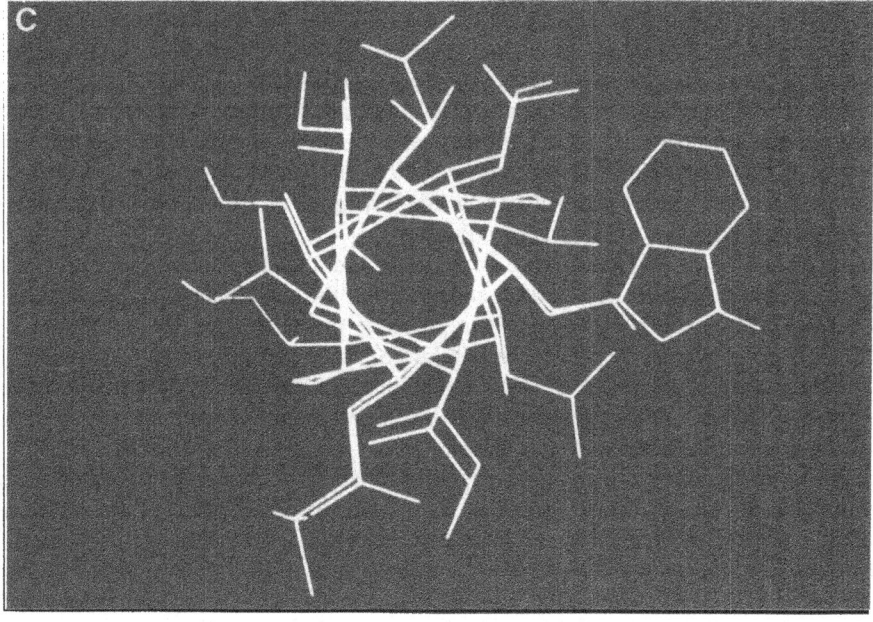

Figure 3 End-on view of the structures of the individual transmembrane α-helices of the chloroplast phosphate translocator. (A), helix 1; (B), helices 2 (left) and 4 (right). The disulfide bridge is indicated by an arrow; (C), helix 3; (D), helices 5 (right) and 6 (left).

82

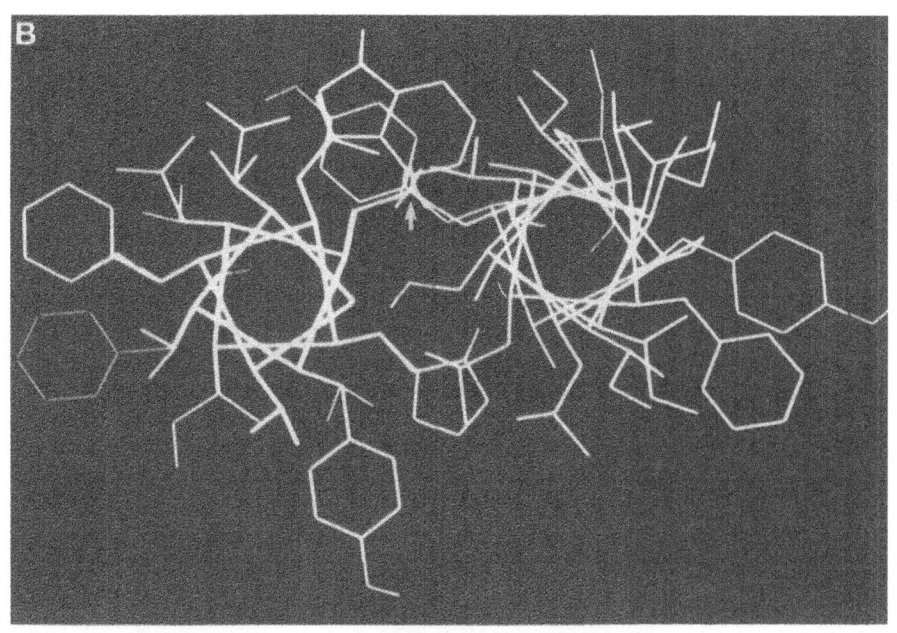

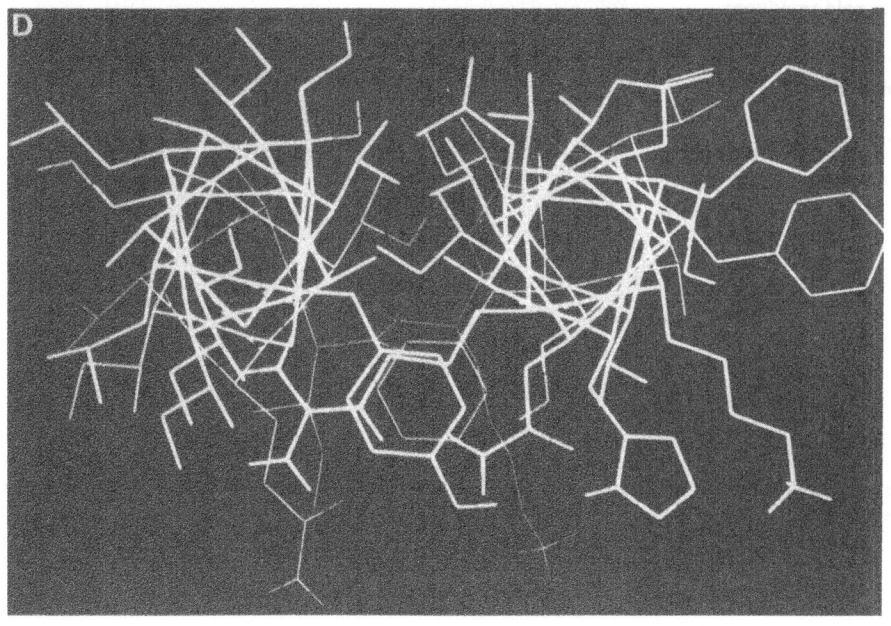

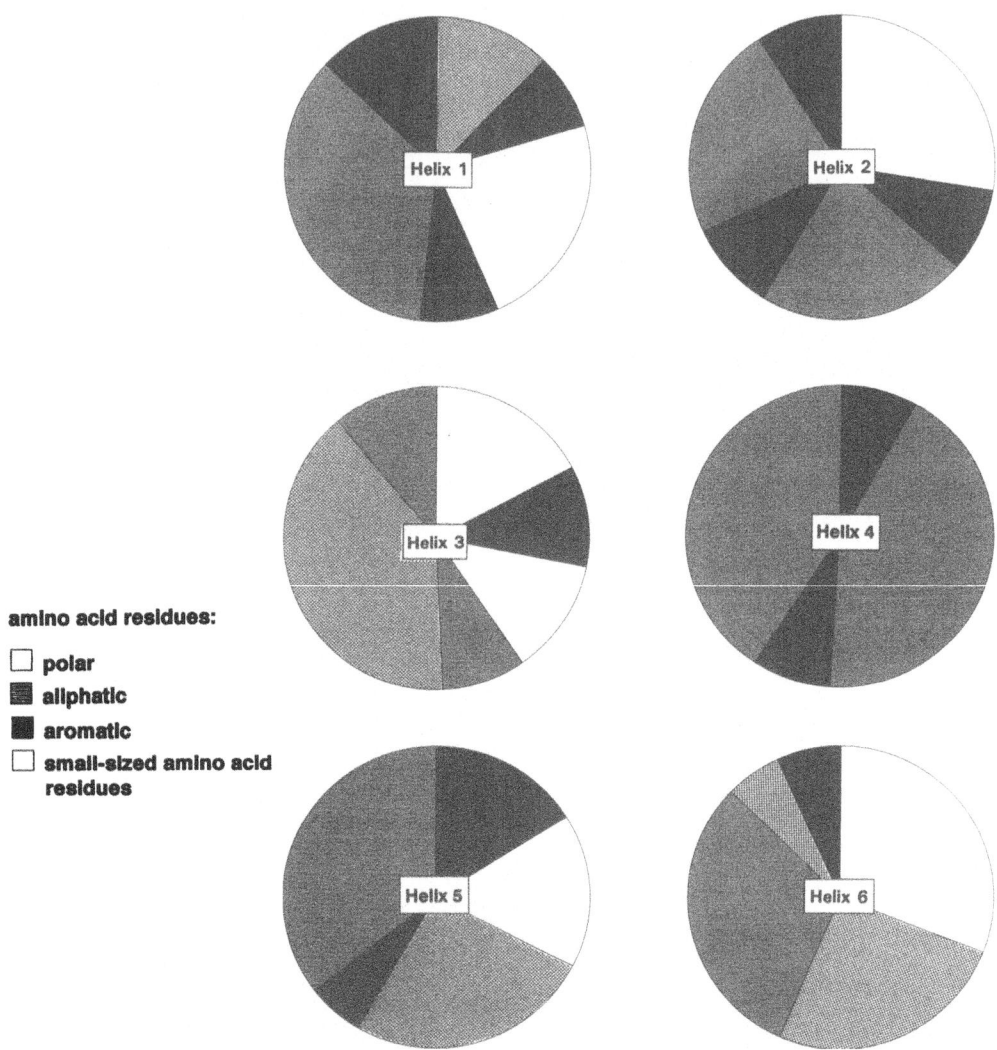

amino acid residues:

- ☐ polar
- ▦ aliphatic
- ■ aromatic
- ☐ small-sized amino acid
  residues

Figure 4 Schematic representation of the distribution of side chains on the surface of the transmembrane ∝-helices 1-6.

84

Furthermore, a hydrogen bond may be formed between Lys_{273} (helix 5) and Tyr_{64} (helix 2):

$$- Phe_{25}$$
$$- Lys_{273} \cdots\cdots\cdots Tyr_{64} -$$
$$- Trp_{29}$$

helix 5   helix 1   helix 2

It is tempting to speculate that these four amino acids could be involved in the relais-type transport mechanism of the phosphate translocator enabling transport in only an antiport manner (as observed under physiological conditions) and preventing transport in only one direction.

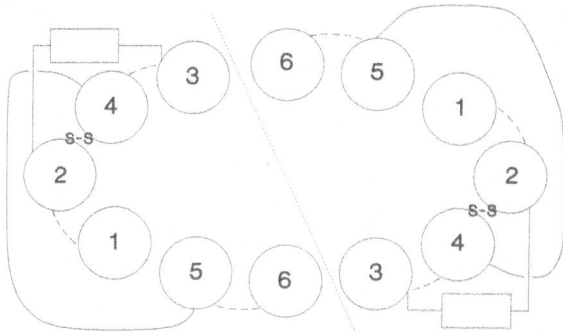

Figure 5 Schematic alignment of the 12 $\propto$-helices in the dimeric phosphate translocator protein. - - -, ————— , loops located below or above the membrane plane, respectively. ••••••, C2-symmetry axis.

3. Concluding Remarks

Although much progress is needed before we shall obtain a physical description of the molecular mechanism of a metabolite transport system, the avaibility of the cloned chloroplast phosphate translocator gene in combination with the methods presented in this paper has opened the way to insights into structure-function relationships of this translocator protein.

Acknowledgement

This work was supported by the Bundesministerium für Forschung und Technologie (BMFT 0319295A).

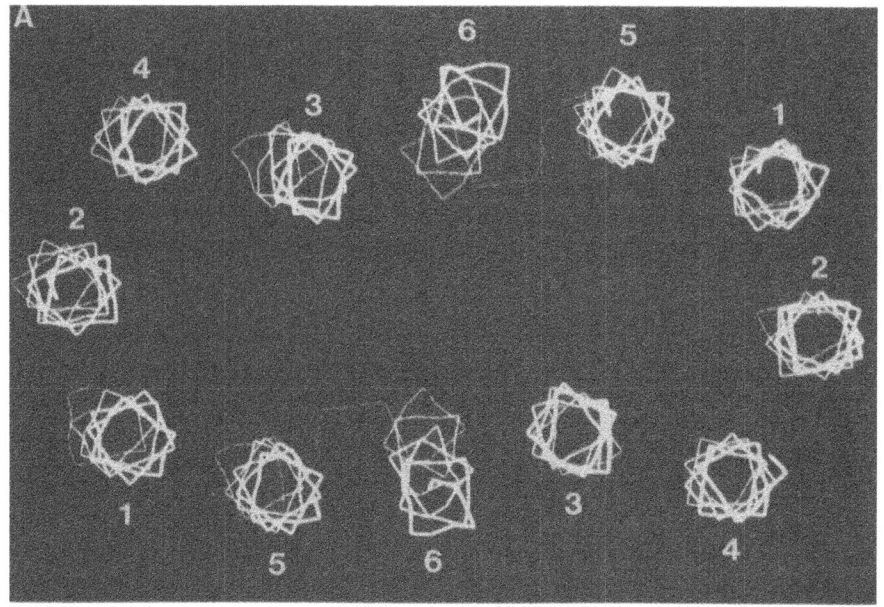

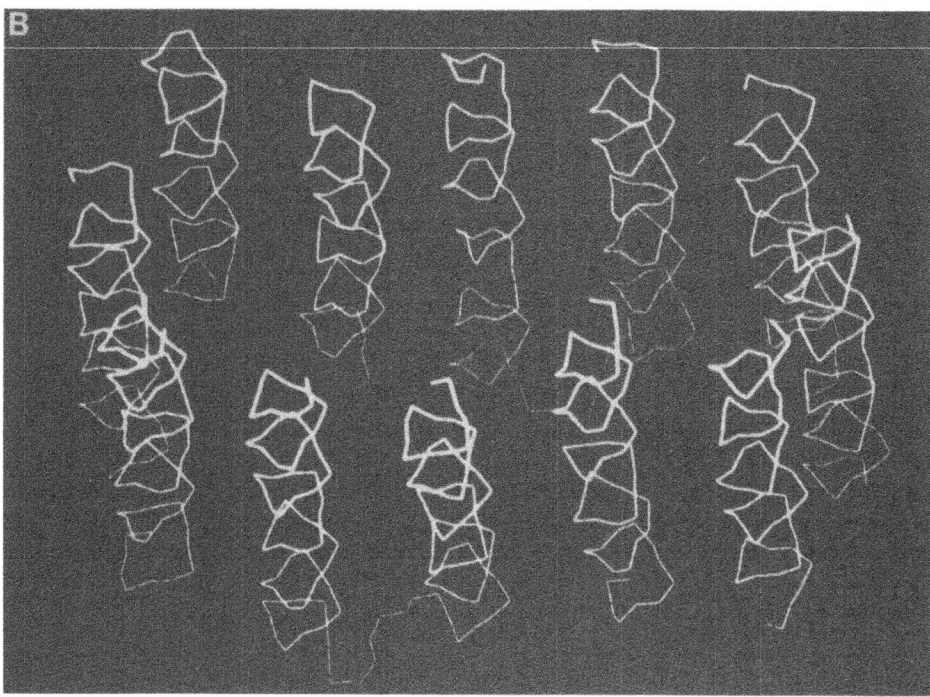

Figure 6 Computer-generated arrangement of the 12 transmembrane helices within the dimeric chloroplast phosphate translocator. (A), top view; the numbers refer to helices 1-6 of each monomer. (B), in side view. Only the backbones are shown.

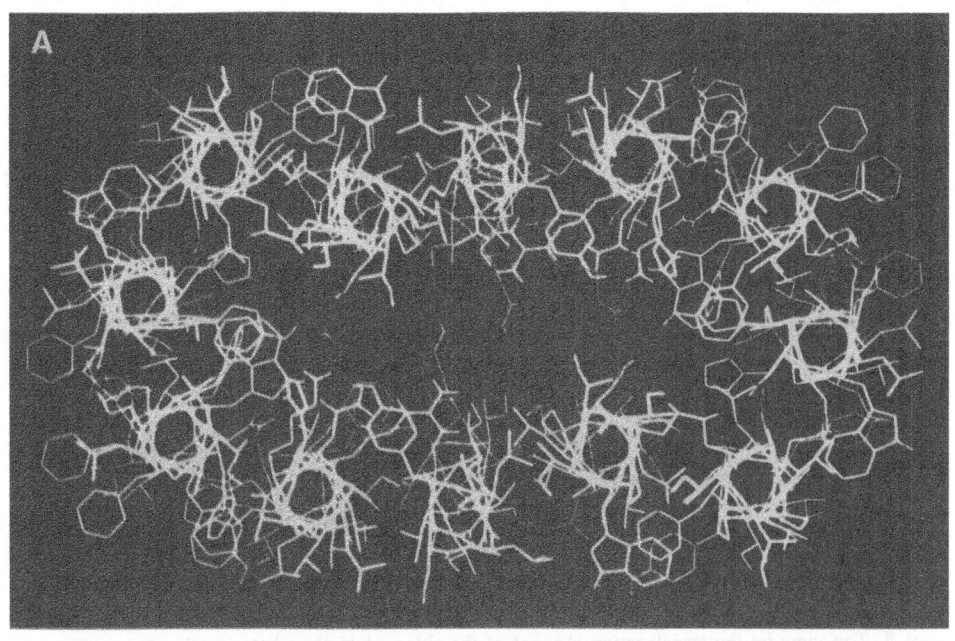

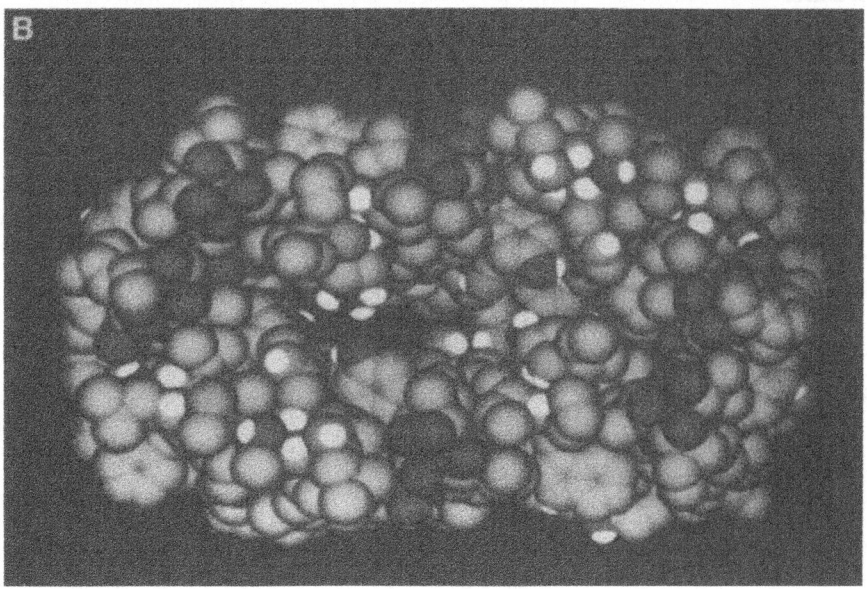

Figure 7 Top view on the computer-generated three-dimensional structure of the chloroplast phosphate translocator. (A), vector representation; (B) space-filling representation.

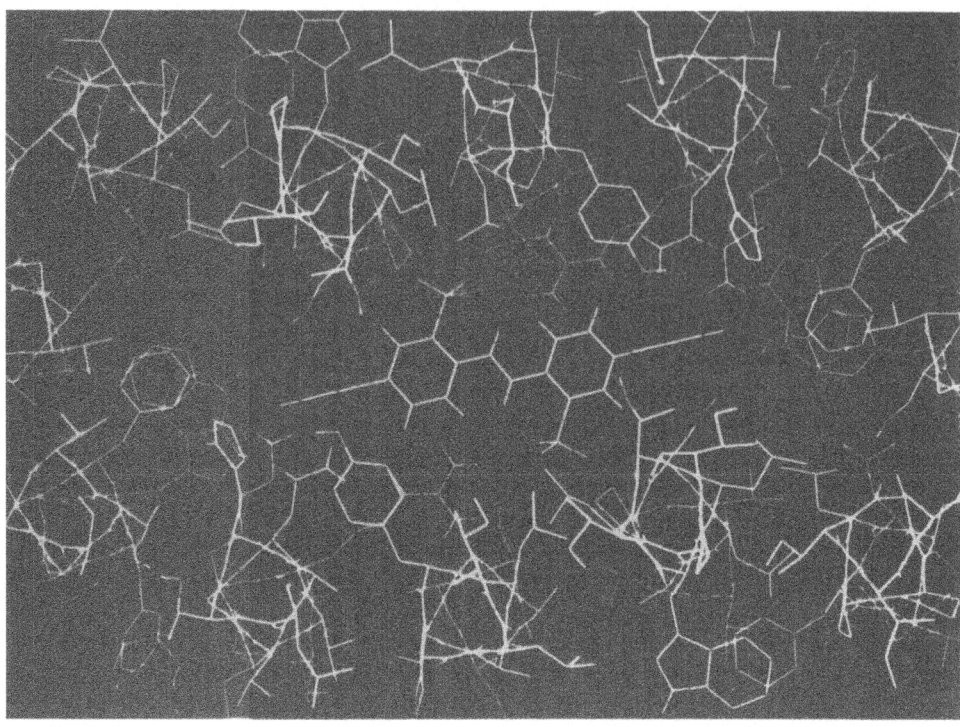

Figure 8 Top view on the translocation channel of the chloroplast phosphate translocator protein with the inhibitor DIDS.

REFERENCES

BORCHERT, S., GROßE, H., and HELDT, H.W., 1989. Specific transport of phosphate, glucose 6-phosphate, dihydroxyacetone phosphate and 3-phosphoglycerate into amyloplasts from pea roots. *FEBS Letters*, **253**, 183-186.

DRESES-WERRINGLOER, U., FISCHER, K., WACHTER, E., LINK, T.A., and FLÜGGE, U.-I., 1990. cDNA sequence of the precursor of the 37 kDa inner envelope membrane polypeptide from spinach chloroplasts: its transit peptide contains an amphiphilic ∝-helix as the only detectable structural element. *European Journal of Biochemistry*, **195**, 361-368.

FLÜGGE, U.-I., and HELDT, H.W., 1979. Phosphate translocator in chloroplasts: Identification of the functional protein and characterization of its binding site. In: *Function and molecular aspects of biomembrane transport*. Eds E. Quagliariello, F. Palmieri, E.M. Klingenberg, Elsevier/North Holland Biomedical Press, Amsterdam. pp 373-382.

FLÜGGE, U.-I., 1985. Hydrodynamic properties of the Triton X-100-solubilized chloroplast phosphate translocator. *Biochimica Biophysica Acta*, **815**, 299-305.

FLÜGGE, U.-I., FISCHER, K., GROSS, A., SEBALD, W., LOTTSPEICH, F., and ECKERSKORN, C., 1989. The triose phosphate-3-phosphoglycerate- phosphate translocator from spinach chloroplasts: nucleotide sequence of a full-length cDNA clone and import of the *in vitro* synthesized precursor protein into chloroplasts. *EMBO Journal*, **8**, 39-46.

FLÜGGE, U.-I., and WESSEL, D., 1984. Cell-free synthesis of putative precursors for envelope membrane polypeptides of spinach chloroplasts. *FEBS Letters*, **168**, 255-259.

FLÜGGE, U.-I., and HELDT, H.W., 1991. Metabolite translocators of the chloroplast envelope. *Annual Reviews of Plant Physiology and Plant Molecular Biology*, **42**, 129-144.

FLÜGGE, U.-I., WEBER, A., FISCHER, K., LOTTSPEICH, F., ECKERSKORN, C., WAEGEMANN, K., and SOLL, J., 1991. The major chloroplast envelope polypeptide is the phosphate translocator and not the protein import receptor. *Nature*, **353**, 364-367.

GROSS, A., BRÜCKNER, G., HELDT, H.W., and FLÜGGE, U.-I., 1990. Comparison of the kinetic properties, inhibition and labelling of the phosphate translocators from maize and spinach mesophyll chloroplasts. *Planta*, **180**, 262-271.

KYTE, J., and DOOLITTLE, R.F., 1982. A simple method for displaying the hydropathic character of a protein. *Journal of Molecular Biology*, **157**, 105-132.

PAIN, D., KANWAR, Y.S., and BLOBEL, G., 1988. Identification of a receptor for protein import into chloroplasts and its localization to envelope contact sites. *Nature*, **331**, 232-237.

RUMPHO, M.E., EDWARDS, G.E., YOUSIF, A.E., and KEEGSTRA, K., 1988. Specific labelling of the phosphate translocator in C3 and C4 mesophyll chloroplasts by tritiated dihydro-DIDS (1,2-ditritio- 1,2- [2,2'-disulfo-4,4'-diisothiocyano]-diphenylethane). *Plant Physiology*, **86**, 1193-1198.

SCHULTZ, G.E., and SCHIRMER, R.H., 1983. Principles of Protein Structure. *Advanced Texts in Chemistry*. Ed C.R. Cantor, Springer Verlag, Berlin, Heidelberg, New York.

SCHNELL, D.J., BLOBEL, G., and PAIN, D., 1990. The chloroplast import receptor is an integral membrane protein of chloroplast envelope contact sites. *The Journal of Cell Biology*, **111**, 1825-1838.

WAGNER, R., APLEY, E.C., GROSS, A., and FLÜGGE, U.-I., 1989. The rotational diffusion of chloroplast phosphate translocator and of lipid molecules in bilayer membranes. *European Journal of Biochemistry*, **182**, 165-173.

WEINER, S.J., KOLLMAN, P.A., CASE, D.A., SINGH, U.C., GHIO, C., ALAGONA, G., PROFETA, S., and WEINER, P., 1984. A new force field for molecular mechanical simulation of nucleic acids and proteins. *Journal of the American Chemical Society*, **106**, 765-784.

WILLEY, D.L., FISCHER, K., WACHTER, E., LINK, T.A., and FLÜGGE, U.-I., 1991. Molecular cloning and structural analysis of the phosphate translocator from pea chloroplasts and its comparison to the spinach phosphate translocator. *Planta*, **183**, 451-461.

Salt-Induced Changes in Ion Transport: Regulation of Primary Pumps and Secondary Transporters

F.M. DuPont

USDA Agricultural Research Service
Western Regional Research Centre
800 Buchanan Street
Albany, CA, USA

1.   Introduction

Research on salt tolerance is a source of challenging questions about the regulation of ion transport in plants. For example, storage of Na^+ in barley (*Hordeum vulgare* L.) vacuoles may involve a complex sequence of events. Within minutes of exposure to NaCl a Na^+/H^+ antiporter is activated, presumably by a signal transduction event (Garbarino and DuPont, 1989).   Several days later, transport by the tonoplast ATPase increases (DuPont and Morrissey, 1991; Matsumoto and Chung, 1988), suggesting that eventually the activity of the primary proton pump is adjusted to the demand created by the secondary active transport. In this paper, recent data on regulation of ion transport in membranes from NaCl-grown plants are reviewed and used as a framework for speculation about the regulation of pumps, symports and antiports in the tonoplast and plasma membrane. The emphasis will be on the tonoplast ATPase and the Na^+/H^+ antiporter.

2.   Role of Ion Transport in NaCl Tolerance

NaCl tolerance clearly requires regulation of ion transport. In the search for "salt tolerance genes", genes for transport proteins and genes for regulation of transport proteins are strong candidates. The degree of NaCl tolerance and strategies for distribution of ions vary between plants, but some generalisation can be made. The ability to maintain a high K^+/Na^+ ratio in the cytoplasm is thought to be a common feature of NaCl-tolerant plants (Wyn Jones and Storey, 1978; Greenway and Munns, 1980). There is a selective distribution of Na^+, Cl and K^+, with exclusion of Na^+ from growing tissues and enrichment of K^+ in meristematic cells and leaf mesophyll cells (Boursier, Lynch, Lauchli, and Epstein, 1987; Wolf, Munns, Tonnet, and Jeschke, 1991). Halophytes utilise Na^+ and Cl^- as osmotica, storing them in the vacuole and some moderately tolerant non-halophytes may do the same, particularly in older

Transport and Receptor Proteins of Plant Membranes
Edited by D.T. Cooke and D.T. Clarkson, Plenum Press, New York, 1992

91

tissues (Binzel, Hess, Bressan, and Hasegawa, 1988; Greenway and Munns, 1980, Huang and Van Steveninck, 1989, Wyn Jones and Storey, 1978). The proteins required for cation selectivity and redistribution of Na^+ and K^+ (Figure 1) include the primary proton pumps, to provide energy for ion transport (Reuveni, Bennett, Bressan, and Hasegawa, 1990; Matsumoto and Chung, 1988; Serrano, 1985); Na^+/H^+ antiports in the plasma membrane, for pumping excess Na^+ out of the cell (Hassidim, Braun, Lerner, and Reinhold, 1990; Jacoby and Teomy, 1989; Katz, Pick, and Avron, 1989; Watad, Pesci, Reinhold, and Lerner, 1986); Na^+/H^+ antiports in the tonoplast, for pumping Na^+ into the vacuole (Blumwald and Poole, 1987; Garbarino and DuPont, 1988; Staal, Maathuis, Elzenga, Overbeek, and Prins, 1991) and cation channels with high selectivity for K^+ over Na^+ (Pantoja, Dainty, and Blumwald, 1989; Maathuis and Prins, 1990). Selective distribution of Cl is also important, but will not be discussed in this article.

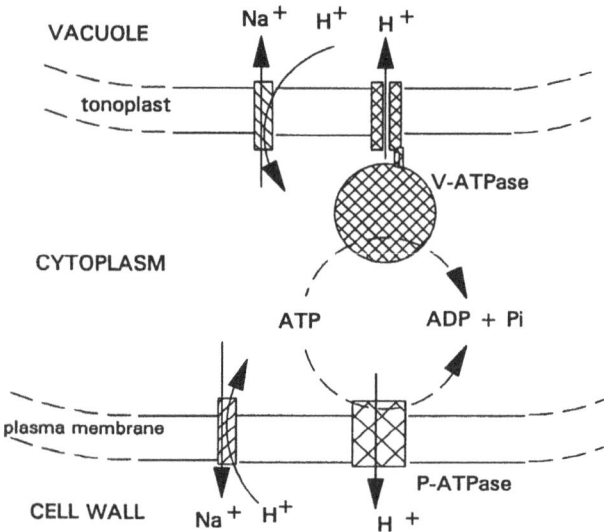

Figure 1 Antiports may pump Na^+ out of the cytoplasm of plant cells, using the pH gradients created by the H^+-ATPases.

Closely related plants, such as wheat (*Triticum* sp.) and barley, may differ in mechanisms for salt tolerance. For example, salt tolerance of various wheat crosses was correlated with a high K^+/Na^+ ratio in the leaves (Gorham, 1990a, Schachtman, Bloom, and Dvorak, 1989), indicating that exclusion of Na^+ from the leaves was a salt-tolerant trait. There was less discrimination between K^+/Na^+ in barley leaves (Gorham, 1990b) perhaps because storage of Na^+ in the vacuole is more important in barley. Differences in the degree of tolerance and mechanism of tolerance may be related to differences in genes for specific ion transport proteins, regulation of gene expression, or regulation of transport activity.

3. Na$^+$/H$^+$ Antiporters

A Na$^+$/H$^+$ antiporter is a transport protein that exchanges Na$^+$ for protons across a membrane (Krulwich, 1983; Grinstein, Rotin, and Mason, 1989) (Figure 1).

Either a proton gradient (ΔpH) or a Na$^+$ gradient ($\Delta\mu$Na$^+$) can serve as the driving force. In plants, Na$^+$/H$^+$ antiporters were detected in isolated tonoplast from red beet, *Beta vulgaris* L. (Blumwald and Poole, 1985), sugar-beet suspension cells (Blumwald and Poole, 1987), *Plantago* roots (Staal *et al.*, 1991) and barley roots (Garbarino and DuPont, 1988) and in intact vacuoles from *Catharanthus* cells (Guern, Mathieu, Kurkdjian, Manigault, Manigault, Gillet, Beloeil, and Lallemand, 1989). Evidence for the activity of a tonoplast Na$^+$/H$^+$ antiporter in the plasma membrane was obtained using plasma membrane vesicles from *Atriplex* and *Gossypium* (Hassidim *et al.*, 1990), excised barley roots and beet slices (Jacoby and Teomy, 1989), intact tobacco (*Nicotiana*) cells (Watad *et al.*, 1986) and the alga, *Dunaliella salina* (Katz *et al.*, 1989).

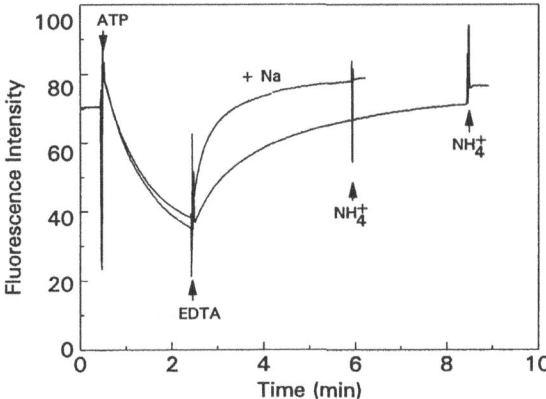

Figure 2 Use of acridine orange to measure the V-ATPase and the Na$^+$/H$^+$ antiport in isolated tonoplast vesicles from barley roots. Vesicles were incubated with 2 μM acridine orange, 0.25 M sorbitol, 10 mM Tris-Mes, pH 7.8, 50 mM choline chloride, 1 mM EGTA and 1 mM MgCl$_2$. ATP (1 mM), 2 mM EDTA or 3 mM EDTA plus 30 mM Na gluconate and 3 mM NH$_4$Cl were added as indicated.

3.1 Measurement and characterisation of Na$^+$/H$^+$ antiporters

The ability to measure Na$^+$/H$^+$ exchange in isolated vesicles makes it possible to characterise the antiport activity in detail. The antiporter can be detected by generating a ΔpH, acid inside, adding Na$^+$ and then measuring either proton efflux (Figure 2) or uptake of ^{22}Na (Table 1) (Barkla, Charuk, Cragoe, and Blumwald, 1990; Garbarino and DuPont, 1988; Katz *et al.*, 1989; Staal, *et al.*, 1991). A ΔpH is generated either by a pH jump technique or by addition of ATP and Mg^{2+} to promote proton transport by the tonoplast ATPase. The ΔpH can be detected by using a fluorescent dye, such as acridine orange (Figure 2).

In the example shown in Figure 2 proton transport was initiated by addition of ATP and Mg to sealed, right-side out tonoplast vesicles isolated from barley roots. As protons were pumped into the vesicles and the pH declined, acridine orange accumulated within the vesicles and the fluorescence of acridine orange decreased. The pump was inactivated by adding EDTA to chelate the Mg, and the slow outward leak of protons was observed as a return of fluorescence to the original level. Addition of Na^+ increased the proton efflux. The relative rate of the antiporter can be expressed as initial rate of relaxation of the quench. A K_m for Na^+ between 5 and 15 mM Na^+ was obtained when the antiporter was measured by this method or by uptake of ^{22}Na (Blumwald, Cragoe, and Poole, 1987; Garbarino and DuPont, 1988).

Table 1 Uptake of ^{22}Na by Membranes Isolated from Barley Roots.
Membranes were added to 5 mM NaCl(1 cpm/pmole of ^{22}Na), 1 mM $MgCl_2$, 1 mM EGTA, 50 mM choline Cl, 0.25 M sorbitol and 5 mM Pipes KOH, pH 7.5 for 10 minutes at 22° C. ATP was 1 mM where included. Data are the average of 2 experiments.

Membrane Fraction	Net Uptake of ^{22}Na		
	-ATP	+ATP	ATP-stimulation
	(nmole Na^+/mg protein)		
Tonoplast from control roots	16	17	1
Tonoplast from salt-grown roots	24	88	64
Plasma membrane from salt-grown roots	9	11	2

The anti-diuretic compound, amiloride, is a potent inhibitor of Na^+/H^+ antiporters and Na^+ channels in animal cells (Grinstein *et al.*, 1989). Amiloride inhibited the Na^+/H^+ antiporter from sugar-beet (Blumwald *et al.*, 1987), but not from barley or *Plantago* (Garbarino and DuPont, 1988; Staal *et al.*, 1991). The amiloride analogue, methyl isobutyl amiloride, inhibited the antiporter from sugar-beet suspension cells with a K_i of approximately 5 μM (Barkla *et al.*, 1990). Even this potent analogue failed to inhibit the Na^+/H^+ antiporter from barley roots, although at concentrations above 2 μM it inhibited transport by the tonoplast ATPase and increased the passive leak (DuPont, unpublished data). However, the antiporter from barley was inhibited by the compounds dodecyl tetraethyl ammonium and dodecyl tetramethyl ammonium (Garbarino and DuPont, 1988).

This laboratory has not been able to detect a Na^+/H^+ antiporter in the plasma membranes from barley roots (Table 1) although high rates of proton transport were measured in the same vesicles. It is possible that the antiporter is most active in the epidermal and root hair cells and is not well represented in the isolated vesicles, or that it is damaged during membrane preparation. If there is such an antiporter in the plasma membrane, it is likely also to be closely regulated.

3.2   Regulation of Na^+/H^+ antiporters

Two forms of up-regulation of the Na^+/H^+ antiporter in response to Na^+ were observed. In barley (Table 1) or *Plantago maritima* the antiport activity was not detected unless the plants were exposed to NaCl (Garbarino and DuPont, 1988; Stall *et al.*, 1991). In sugar-beet membranes, the activity was constitutive, but growth in NaCl increased the antiport activity (Blumwald and Poole, 1987). The antiporter must be a constitutive protein in barley, since activation was very rapid and was complete within 30 minutes, even if protein synthesis was completely inhibited (Garbarino and DuPont, 1989). The time course and sensitivity to inhibitors of protein synthesis for activation of the antiporters from *Plantago* or sugar-beet is not known. Rapid activation of the antiporter in barley made it easy to study the specificity of the eliciting signal for activation of the antiporter (Garbarino and DuPont 1989). The primary signal appeared to be Na^+, since only exposure to Na^+ salts elicited the response. There was no effect of an osmotic shock or exposure to K^+, and Cl^- was not required. Growth in NaCl did not alter uptake of Ca^{2+} by the same membranes (DuPont, unpublished data).

The mechanism of activation of the antiporter is unknown. Comparison with the Na^+/H^+ antiporter in animal plasma membranes may be relevant. The antiporter is activated rapidly when animal cells are exposed to hypo-osmotic shock or if the cytoplasm is acidified (Grinstein *et al.*, 1989; Sardet, Franchi, and Pouyssegur, 1989). In either case the cells take up the osmotically active ion, Na^+, in exchange for the buffered, osmotically inactive H^+ and alkalinise the cytoplasm. Subsequently, a series of other transport systems, including the Na^+/K^+ ATPase, act to restore the balance of cytoplasmic ions and pH. The antiporter is also activated by various mitogenic compounds (Grinstein *et al.*, 1988). There is much evidence for a complex signal transduction system between the initial stimulus and the activation of the antiporter, and G proteins and protein kinases are involved (Sardet *et al.*, 1989). Research on the exact mechanism of activation of the antiporter in animal cells has been greatly retarded by the difficulty of identifying and purifying the antiporter protein, although at least one gene sequence is published (Sardet *et al.*, 1989). There are also gene sequences for a bacterial Na^+/H^+ antiporter, but it seems to be different from the antiporter in animal cells (Taglicht, Padan and Schuldiner, 1991).

There is some hope of identifying the protein for the plant vacuolar Na^+/H^+ antiporter. It was proposed that a 170 kDa polypeptide, labelled with ^3H methyl isobutyl amiloride, might be a component of the Na^+/H^+ antiporter of sugar-beet vacuoles (Barkla *et al.*, 1990). When sugar-beet cells were grown in NaCl, an increase in the 170 kDa polypeptide band was visible on 1D SDS PAGE after staining with Coomassie blue. An antibody against this polypeptide inhibited the Na^+/H^+ antiport activity (Blumwald and Barkla, personal communication). However, no change in amount of any integral membrane protein that was specific to the tonoplast fraction was detected when membranes from control and salt-grown barley roots were compared by 1D and 2D-SDS polyacrilamide gel electrophoresis (Hurkman, Tanaka and DuPont, 1988). Also, no change in phosphorylation of any tonoplast polypeptides was observed when roots were treated with 100 mM NaCl (Garbarino, Hurkman, Tanaka, and DuPont, 1991).

3.3   The role of the antiporter in salt tolerance

It is clear that a Na^+/H^+ antiporter that pumps Na^+ into the vacuole may be a useful feature for a salt-tolerant plant. Why it should be a constitutive protein that is activated by salt is not so clear. If rapid activation of the antiporter is a useful trait, it suggests that barley roots may suddenly be exposed to Na^+ when the plants are growing in the field. For example, if Na^+ is unevenly distributed in the soil, roots may suddenly be exposed to excess Na^+ when rainfall washes Na^+ past the roots. It also remains to be demonstrated that the antiporter is essential to salt tolerance and to what degree the salt tolerance of a particular plant can be attributed to the presence or absence of this protein. The antiporter was detected only in membranes from the salt tolerant *Plantago maritima*, and not in the salt sensitive species, *Plantago media*, (Stall *et al.*, 1991), so there is at least a correlation between the activity and salt tolerance. Since it is likely that salt tolerance is not a single gene phenomenon, proving a specific relationship between the antiporter and salt tolerance may be complicated.

4.   Regulation of ATPases in Response to NaCl

The ΔpH created by the tonoplast and plasma membrane H^+-ATPases provides the driving force for the Na^+/H^+ antiporters in plant cells. When the plant is exposed to salt, ATPase activities may be modified in order to maintain or increase the ΔpH across the tonoplast and plasma membrane and to maintain the cytoplasmic pH balance. There might be immediate up- or down-regulation of ATPase activity by a regulatory factor, or long range changes that involve gene expression and protein synthesis. These changes may be tissue specific. Many techniques are available to investigate the role of the ATPases in salt tolerance. Membranes can be isolated and examined for changes in ATPase activity (Matusumoto and Chung, 1988; Reuveni *et al.*, 1990; Staal *et al.*, 1990) and cDNA probes for ATPase polypeptides can be used to detect changes in expression of ATPase genes. Characterisation of the ATPase genes and polypeptides might reveal that the activity is modified by replacing one isozyme with another (Boutry, Michelet, and Goffeau, 1989; Ewing, Wimmers, Meyer, Chetelat and Bennett, 1990; Harper, Manney, DeWitt, Yoo, and Sussman, 1990). Regulation by cytoplasmic factors, such as protein kinases, specific lipids or small molecules may be detected with the appropriate experimental techniques (Briskin, 1990; Palmgren, Larsson, and Sommarin, 1990; Sanders, 1990). However, to calculate the outcome of these various scenarios it is useful to consider the bioenergetics of ATPases and antiporters. Increasing the number of ATPases will have a different consequence from altering the turnover rate or changing the net stoichiometry of ATP hydrolysed to protons pumped.

4.1   Regulation of the ATPase to increase ΔpH

An increase in ΔpH will increase the driving force for uptake of Na^+ into the vacuole or for expulsion of Na^+ from the cytoplasm. The driving force, $\Delta\mu Na^+$, is directly related to the size of the ΔpH (Johnson, Carty, and Scarpa, 1981):

$$\Delta\mu Na^+ = (-2.3RT/F) \bullet \Delta pH = (-2.3RT/F)\log[Na_i]/[Na_o] \qquad (1)$$

The upper limit for the ΔpH is determined by the electrochemical gradient for protons, $\Delta\bar{\mu}H^+$, which is limited by the energy available in the hydrolysis of ATP, ΔG_{ATP}, not by the number of ATPases. Let Δp be the proton potential, at equilibrium; let $\Delta\psi$ = membrane potential; let n be the ratio of protons transported to ATP hydrolysed and let ΔG = the free energy required to pump a proton against the proton potential (Harold, 1986).

$$\Delta p = \Delta\bar{\mu}H^+/F = \Delta\psi - 59 \ \Delta pH \ (mV) \tag{2}$$

$$\Delta p = \Delta G/nF \tag{3}$$

The upper limit for $\Delta\bar{\mu}H^+$ can be increased only by decreasing n, the stoichiometry for coupling of ATP hydrolysis to proton transport. The upper limit for ΔpH can be increased by decreasing n or decreasing $\Delta\psi$. If it is necessary to increase the driving force for Na^+ uptake, one might imaging a scenario whereby the ATPase is modified in some way that decreases n, or Cl^- permeability might be increased, to decrease $\Delta\psi$. Increasing the number of ATPases will not increase the size of the ΔpH when the ATPase is pumping into a limited space, such as the vacuole. Increasing the number of proton pumps might be an effective strategy to increase ΔpH at the plasma membrane where the pumps are transporting ions into a void (Nelson, 1991).

It may not be necessary to increase ΔpH to drive the Na^+/H^+ antiporter. For example, a gradient of 3.8 pH units is within the capacity of an ATPase with a stoichiometry of 2 H^+/ATP, which is the current estimate for n for the tonoplast ATPase (Rea and Sanders, 1987). Similarly, 2 pH units are sufficient to support a Na^+ gradient with $[Na^+]_i/[Na^+]_o = 100$ (Equation 1). Nelson (1991) suggested that vacuolar type ATPases in endomembrane compartments are not limited by thermodynamic constraints, that other factors maintain the ΔpH below the theoretical limit and that the coupling efficiency of proton ATPases in endomembrane compartments is regulated by slippage, or intrinsic uncoupling.

4.2 Increasing the rate of proton transport

A high rate of Na^+/H^+ exchange across the tonoplast or plasma membrane may require a concomitant increase in the rate of H^+ transport by the H^+-ATPase, rather than an increase in ΔpH. This could be accomplished by increasing the number of pumps, increasing n, or increasing the turnover number, or modulating the degree of slippage.

There is a small but growing amount of data that suggest the activity of the tonoplast ATPases is modified when plants are exposed to NaCl. The specific activity for proton transport increased in salt-grown, salt-adapted *Nicotiana* cells, based on the amount of ATPase protein detected with antibodies (Reuveni *et al.*, 1990). Exposure to 200 mM NaCl for several days resulted in a 2- or more fold increase in proton transport by the tonoplast ATPase from barley roots (Matusumoto and Chung, 1988). However, it was reported that exposing roots to 100 mM NaCl had no effect on the rate of ATP hydrolysis for the tonoplast ATPase of barley roots (Garbarino and DuPont, 1988) or *Plantago maritima* (Staal *et al.*, 1991).

Recently, we made a detailed examination of the effect of growth in NaCl on the ATP hydrolysis and proton transport by the tonoplast ATPase from barley roots. There was a 2-fold increase in the specific activity of proton transport (Figure 3), but no change in ATP hydrolysis or in the amount of ATPase polypeptides detected with antibodies (DuPont and Morrissey, 1991). The increase in proton transport required several days exposure to NaCl (Figure 3) and Matusumoto and Chung (1988). Currently, we are working on the hypothesis that growth in NaCl leads to a change in turnover, stoichiometry or coupling for the tonoplast ATPase from barley roots, which results in an increased supply of protons per ATP hydrolysed. The time scale suggests that gene expression, protein synthesis and synthesis of new ATPase molecules are required for this change in the ATPase activity. A similar phenomenon may occur in salt-adapted tobacco cells (Reuveni *et al.*, 1990).

There is only one preliminary report on the effects of salt on expression of genes for the tonoplast ATPase (Hasegawa, Binzel, Reuveni, Watad, and Bressan, 1989) and none on the plasma membrane ATPase. It is unlikely that there are large increases in gene expression, on the scale of the 100 or 1000 fold increases that make for easily observed changes in mRNAs and protein synthesis. However, there was a 6-fold increase in message for the plasma membrane ATPase in drought stressed soybean (*Glycine max* L.) (Surowy and Boyer, 1991), and it will be interesting to learn whether similar changes can be detected in response to salt. It may be necessary to use specific cDNA probes for individual ATPase isozymes to explore the role of the ATPases in the selected distribution of Na^+, Cl^- and K^+ between different plant tissues.

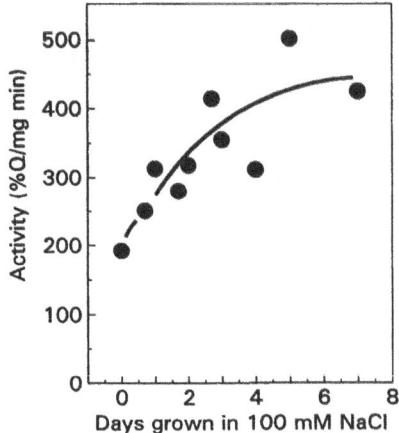

Figure 3 Proton transport as a function of the time that roots were exposed to 100 mM NaCl. Roots were germinated and grown over nutrient medium for 7 days and all roots were harvested on day 7. Membranes were isolated and transport was assayed as initial rate of quench of acridine orange fluorescence.

5.   Summary

Exposing plants to environmental challenges such as excess NaCl can elicit changes in ion transport that can be observed in isolated membrane vesicles. These isolated membranes are a valuable experimental system for studying the regulation of ion transport. By comparing the ATPases and membrane vesicles from control and salt-treated plants much may be learned about how pumps and antiporters are regulated when plants are challenged with high concentrations of NaCl. Recently, it also has become possible to look for the effects of salt on expression of the genes for ion transport proteins. Such studies will provide different and complementary information on the role of ion transport in salt tolerance.

Acknowledgement

MIA was a gift from E. Blumwald.

REFERENCES

BARKLA, B.J., CHARUK, J.H.M., CRAGOE, E.J.JR., and BLUMWALD, E., 1990. Photolabelling of tonoplast from sugar-beet cell suspension by [^{3}H]5-(N-methyl-N-isobutyl)-amiloride, an inhibitor of the vacuolar Na$^+$/H$^+$ antiport. *Plant Physiology*, **93**, 924-930.

BINZEL, M.A., HESS, F.D., BRESSAN, R.A., and HASEGAWA, P.M., 1988. Intracellular compartmentation of ions in salt adapted tobacco cells. *Plant Physiology*, **84**, 607-614.

BLUMWALD, E., CRAGOE, E.J., and POOLE, R.J., 1987. Inhibition of Na$^+$/H$^+$ antiport activity in sugar-beet tonoplast by analogs of amiloride. *Plant Physiology*, **85**, 30-33.

BLUMWALD, E., POOLE, R.J., 1985. Na$^+$/H$^+$ antiport in isolated tonoplast vesicles from storage tissue of *Beta vulgaris*. *Plant Physiology*, **78**, 163-167.

BLUMWALD, E., POOLE, R.J., 1987. Salt tolerance in suspension cultures of sugar-beet. *Plant Physiology*, **83**, 884-887.

BOURSIER, P., LYNCH, J., LAUCHLI, A., and EPSTEIN, E., 1987. Chloride partitioning in leaves of salt-stressed sorghum, maize, wheat and barley. *Australian Journal of Plant Physiology*, **14**, 463-473.

BOUTRY, M., MICHELET, B., and GOFFEAU, A., 1989. Molecular cloning of a family of plant genes encoding a protein homologous to plasma membrane H$^+$-translocating ATPases. *Biochemical and Biophysical Research Communications*, **162**, 567-574.

BRISKIN, D.P., 1990. The plasma membrane H$^+$-ATPase of higher plant cells, biochemistry and transport function. *Biochimica et Biophysica Acta*, **1019**, 95-109.

DUPONT, F.M., and MORRISSEY, P.J., 1991. Purification of a vacuolar ATPase from barley roots. *Plant Physiology*, **96**, 13 (Abstract).

EWING, N.N., WIMMERS, L.E., MEYER, D.J., CHETELAT, R.T., and BENNETT, A.B., 1990. Molecular cloning of tomato plasma membrane H$^+$-ATPase. *Plant Physiology*, **94**, 1874-1881.

FAN, T.W.-M., HIGASHI, R.M., NORLYN, J., and EPSTEIN, 1989. *In vivo* ^{23}Na and ^{31}P NMR measurement of a tonoplast Na$^+$/H$^+$ exchange process and its characteristics in two barley cultivars. *Proceedings of the National Academy of Sciences, USA*, **86**, 9856-9860.

GARBARINO, J., and DUPONT, F.M., 1988. NaCl induces a Na$^+$/H$^+$ antiport in tonoplast vesicles from barley roots. *Plant Physiology*, **86**, 231-236.

GARBARINO, J., DUPONT, F.M., 1989. Rapid induction of Na$^+$/H$^+$ exchange activity in barley root tonoplast. *Plant Physiology*, **89**, 1-4.

GARBARINO, J., HURKMAN, W.J., TANAKA, C.K., and DUPONT, F.M., 1991. *In vitro* and *in vivo* phosphorylation of polypeptides in plasma membrane and tonoplast-enriched fractions from barley roots. *Plant Physiology*, **95**, 1219-1228.

GORHAM, J., 1990a. Salt tolerance in the Triticeae: K/Na discrimination in synthetic hexaploid wheats. *Journal of Experimental Botany*, **41**, 623-627.

GORHAM, J., 1990b. Salt tolerance in the Triticeae: ion discrimination in rye and triticale. *Journal of Experimental Botany*, **41**, 609-614.

GREENWAY, H., and MUNNS, R., 1980. Mechanisms of salt tolerance in nonhalophytes. *Annual Review of Plant Physiology*, **31**, 149-190.

GRINSTEIN, S., ROTIN, D., and MASON, M.J., 1989. Na$^+$/H$^+$ exchange and growth factor-induced cytosolic pH changes. Role in cellular proliferation. *Biochimica et Biophysica Acta*, **988**, 73-97.

GUERN, J., MATHIEU, Y., KURKDJIAN, A., MANIGAULT, P., MANIGAULT, J., GILLET, B., BELOEIL, J.-C., and LALLEMAND, J.-Y., 1989. Regulation of vacuolar pH of plant cells. II. A ^{31}P NMR study of the modifications of vacuolar pH in isolated vacuoles induced by proton pumping and cation/H$^+$ exchanges. *Plant Physiology*, **89**, 27-36.

HAROLD, F.M., 1986. A study of bioenergetics. W.H. Freeman and Company, New York.

HARPER, J.F., MANNEY, L., DEWITT, N.D., YOO, M.H., and SUSSMAN, M.R., 1990. The *Arabidopsis thaliana* plasma membrane H$^+$-ATPase multigene family. *Journal of Biological Chemistry*, **265**, 13601-13608.

HASEGAWA, P.M., BINZEL, M.L., REUVENI, M., WATAD, A.A., and BRESSAN, R.A., 1989. Physiological and Molecular Mechanisms of Ion accumulation and Compartmentation Contributing to Salt Adaption of Plant Cells. In *Horticultural Biotechnology*. Eds A.B. Bennett and S.D. O'Neill. Wiley-Liss, New York.

HASSIDIM, M., BRAUN, Y., LERNER, H.R., and REINHOLD, L., 1990. Na$^+$/H$^+$ and K$^+$/H$^+$ antiport in root membrane vesicles isolated from the halophyte Atriplex and the glycophyte cotton. *Plant Physiology*, **94**, 1795-1801.

HUANG, C.X., and VAN STEVENINCK, R.F.M., 1989. Maintenance of low Cl-concentrations in mesophyll cells of leaf blades of barley seedlings exposed to salt stress. *Plant Physiology*, **90**, 1440-1443.

HURKMAN, W.J., TANAKA, C.K., and DUPONT, F.M., 1988. The effects of salt stress on polypeptides in membrane fractions from barley roots. *Plant Physiology*, **88**, 1263-1273.

JACOBY, B., and TEOMY, S., 1988. Assessment of Na^+/H^+ antiport in ATP-depleted red beet slices and barley roots. *Plant Science*, **55**, 103-106.

JOHNSON, R.G., CARTY, S.E., and SCARPA, A., 1981. Proton substrate stoichiometries during active transport of biogenic amines in chromaffin ghosts. *Journal of Biological Chemistry*, **256**, 5773-5780.

KATZ, A., PICK, U., and AVRON, M., 1989. Characterisation and reconstitution of the Na^+/H^+ antiport from the plasma membrane of the halotolerant alga *Dunaliella*. *Biochimica et Biophysica Acta*, **983**, 9-14.

KRULWICH, R.A., 1983. Na^+/H^+ antiporters. *Biochimica et Biophysica Acta*, **726**, 245-264.

MAATHUIS, F.J.M., and PRINS, H.B.A., 1990. Patch clamp studies on root cell vacuoles of a salt-tolerance and a salt-sensitive *Plantago* species. *Plant Physiology*, **92**, 23-38.

MATSUMOTO, H., and CHUNG, G.C., 1988. Increase in proton-transport activity of tonoplast vesicles as an adaptive response of barley roots to NaCl stress. *Plant Cell Physiology*, **29**, 1133-1140.

NELSON, N., 1991. Structure and pharmacology of the proton-ATPases. *Trends in Pharmacological Sciences*, **33**, 71-74.

PALMGREN, M.G., LARSSON, C., and SOMMARIN, M., 1990. Proteolytic activation of the plant plasma membrane H^+-ATPase by removal of a terminal segment. *Journal of Biological Chemistry*, **265**, 13423-13426.

PANTOJA, O., DAINTY, J., and BLUMWALD, E., 1989. Ion channels in vacuoles from halophytes and glycophytes. *Federation of European Biologists Letters*, **255**, 92-96.

REA, P.A., and SANDERS, D., 1987. Tonoplast energisation: Two H^+ pumps, one membrane. *Physiologia Plantarum*, **71**, 131-141.

REUVENI, M., BENNETT, A.B., BRESSAN, R.A., and HASEGAWA, P.M., 1990. Enhanced H^+ transport capacity and ATP hydrolysis activity of the tonoplast H^+-ATPase after NaCl adaptation. *Plant Physiology*, **94**, 524-530.

SANDERS, D., 1990. Kinetic modelling of plant and fungal membrane transport systems. *Annual Review of Plant Physiology and Plant Molecular Biology*, **41**, 77-107.

SARDET, C., FRANCHI, A., and POUYSSEGUR, J., 1989. Molecular cloning, primary structure and expression of the human growth factor-activatable Na^+/H^+ antiporter. *Cell*, **56**, 271-280.

SCHACHTMAN, D.P., BLOOM, A.J., and DVORAK, J., 1989. Salt-tolerant Triticum X Lophopyrum derivatives limit the accumulation of sodium and chloride ions under saline-stress. *Plant, Cell and Environment*, **12**, 47-55.

SERRANO, R., 1985. Plasma membrane ATPase of Plants and Fungi. CRC Press, Inc., Boca Raton, Florida.

STAAL, M., MAATHUIS, F.J.M., ELZENGA, J.T., OVERBEEK, J.H.M., and PRINS, H.B.A., 1991. Na^+/H^+ antiport activity in tonoplast vesicles from roots of the salt-tolerant *Plantago maritima* and the salt-sensitive *Plantago media*. *Physiologia Plantarum*, **82**, 179-184.

SUROWY, T.K., and BOYER, J.S., 1991. Low water potentials affect expression of genes encoding vegatative storage proteins and plasma membrane proton ATPase in soybean. *Plant Molecular Biology*, **16**, 251-262.

TAGLICHT, D., PADAN, E., and SCHULDINER, S., 1991. Overproduction and purification of a functional Na^+/H^+ antiporter coded by nhaA (ant) from *Escherichia coli*. *Journal of Biological Chemistry*, **266**, 11289-11294.

WATAD, A.A., PESCI, P.-A., REINHOLD, L., and LERNER, H.R., 1986. Proton fluxes as a response to external salinity in wild type and NaCl-adapted *Nicotiana* cell lines. *Plant Physiology*, **81**, 454-459.

WOLF, O., MUNNS, R., TONNET, M.L., and JESCHKE, W.D., 1991. The role of the stem in the partitioning of Na^+ and K^+ in salt-treated barley. *Journal of Experimental Botany*, **42**, 697-704.

WYN JONES, R.G., and STOREY, R., 1978. Salt stress and comparative physiology in the Gramineae. IV. Comparison of salt stress in *Spartina x townsendii* and three barley cultivars. *Australian Journal of Plant Physiology*, **5**, 839-850.

Uptake of Malate and Citrate into Plant Vacuoles

E. Martinoia and D. Rentsch

Institute of Plant Sciences
Swiss Federal Institute of Technology
ETH-Zentrum
Sonneggstrasse 5
CH-8092 Zürich, Switzerland

Of the different organic anions which are often present a high concentrations in plants, malate plays a central role. Plants exhibiting crassulacean acid metabolism (CAM) fix CO_2 with the enzyme phosphoenolpyruvate carboxylase during the night and accumulate large amounts of malic acid. During the light period, malic acid is decarboxylated and the released CO_2 is fixed in the Calvin cycle. C4 plants fix CO_2 in the mesophyll in a similar reaction during the day, as CAM in the dark. In these plants, malate is transferred to the bundle sheaths, decarboxylated and the CO_2 fixed in the photosynthetic reaction. This reaction enables the plant to fix CO_2 more efficiently, since the affinity of phosphoenolpyruvate carboxylase to HCO_3^- is much higher than that of ribulose-1,5- diphosphate carboxylase to CO_2. Diurnal fluctuations of malate can also be observed in C3 plants. However, in these plants malate is accumulated during the day and used as an energy source for respiration in the dark (Winter, Usuda, Tsuzuki, Schmitt, Edwards, Thomas, and Evert, 1982; Gerhardt, Stitt, and Heldt, 1987). Malate metabolism and accumulation also play an important role during the opening of stomata since, in most plants, malate is used for balancing K^+ (Schnabl and Kottmeier, 1984). Other prominent organic acids often accumulated at high concentrations in plants include shikimic acid, which is present mainly in gymnosperms and some woody angiosperms, as well as gallic, oxalic and citric acid.

Due to the central role of malate in plant metabolism, investigations of transport processes involving organic acids have focused on malate. Osmotic considerations suggest that malate, which may accumulate up to 300 mM during the night in CAM plants, is localised in the large central vacuole. Compartmentation studies confirmed that almost all of the cellular malate is localised in the vacuole (Buser and Matile, 1977). Using the non-aqueous fractionation method, Gerhardt *et al.* (1987) could demonstrate that malate in C3 plants is also mainly localised in the central vacuole and that fluctuations in the malate content could be attributed to vacuolar fluctuations, whereas the cytosolic malate content

Transport and Receptor Proteins of Plant Membranes
Edited by D.T. Cooke and D.T. Clarkson, Plenum Press, New York, 1992

101

remained constant. In this case too, the vacuolar malate concentration was consistently higher than the cytosolic, indicating the involvement of the tonoplast ATPase or PPase (Sze, 1985; Rea and Sanders, 1987). Malate is one of the main products in photosynthesising protoplasts from barley leaves (Kaiser, Martinoia, and Wiemken, 1982). Compartmentation analysis, using a very fast vacuole isolation procedure, showed that malate is transferred very rapidly to the vacuole (Figure 1), indicating that a transport system is involved.

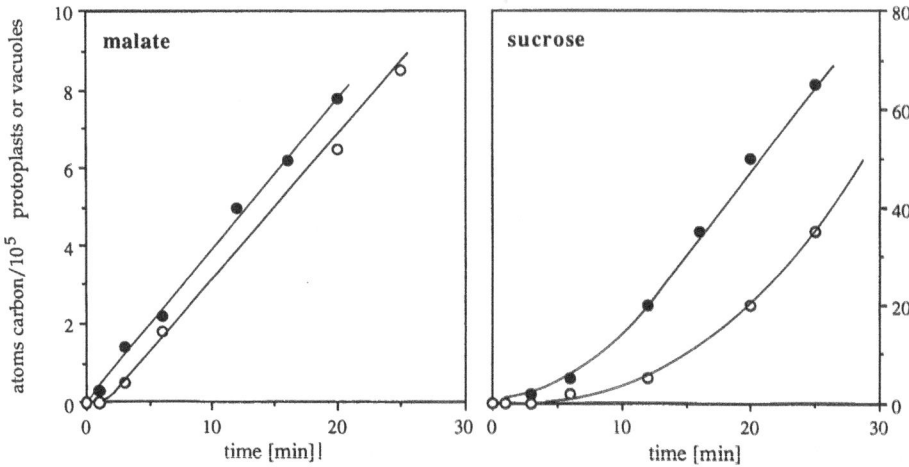

Figure 1 Time-dependent appearance of newly synthesised malate and sucrose in protoplasts (●) and vacuoles (○). Data from Kaiser *et al.*, (1982).

A permease for malic acid was first described by Buser-Sutter, Wiemken, and Matile (1981), using vacuoles from the CAM plant *Kalanchoe daigremontiana*. The apparent K_m for malic acid was about 1 mM; the uptake system was not specific for malic acid since other dicarboxylic acids were inhibitory. Contrary to expectations, the authors did not detect an energy dependency of malic acid uptake, which would seem to be necessary for thermodynamic reasons (Lüttge, Smith, Marigo, and Osmond, 1981). The observed uptake rates were too low to explain the fluxes of malate observed *in vivo*. MgATP-dependent malic acid uptake into barley mesophyll vacuoles was observed by Martinoia, Flügge, Kaiser, Heber, and Heldt (1985). Determination of the vacuolar malic acid concentration showed that the ATP-stimulated uptake occurred against a concentration gradient and was not the result of enhanced malic acid exchange. An accumulation of malate within the vacuole could be observed even in the absence of ATP. Addition of MgATP did not change the K_m (1.6 to 3 mM) but resulted in an enhanced V_{max}. Similar observations were made by Nishida and Tominaga (1987) and Marigo, Bouyssou, and Laborie (1988). Interestingly, in CAM plants malate is more effective than chloride in stimulating ATP and PPi-dependent vesicle acidification (White and Smith, 1989; Marquardt-Jarczyk and Lüttge, 1990). This can be taken as a measure of transport efficiency. In C3 plants the effect of malate was much lower than that of chloride (Kästner and Sze, 1987; Pope and Leigh, 1987). Surprisingly, fumarate was even more effective that malate in stimulating ATP-dependent vesicle acidification or dissipation of a pre-existing membrane potential in *Kalanchoe* (White and Smith, 1989).

Table 1 Properties of the malate carriers from different plants.

	Hordeum Vulgare (barley)[a,b,c,d]	Catharanthus[e,f]	Kalanchoe daigremontiana[g,h,i]
K_m	1.0 - 3.2 mM	2 (pH 5.5) to 20 (pH 7.5) mM	1.0 to 10 mM
Inhibition of malate transport by carboxylic acids (in parentheses mean K_i where determined)	citrate (1 mM) > benzenetri-carboxylate (1.2 mM) = phenylsuccinate = D-malate > tartronate (5 mM) > succinate = fumarate > hydroxybutyrate	D-malate > oxaloacetate > fumarate = succinate > malonate = citrate > shikimate > aconitate = isocitrate	D-malate = tatrtronate = phenylsuccinate > citrate = oxalacetic acid > succinate = fumarate
Inhibitors (concentration) % inhibition	DIDS (50 - 200 μM) 85-90% Pyridoxal phosphate (200 - 500 μM) 60-90% Diethylpyrocarbonate (1mM) 70% PCMBS (1mM) 98%	DIDS (200 μM) 70% Dansyl chloride (200 μM) 52% Diethylpyrocarbonate (1 mM) 65%	$HgCl_2$ (mM) 100%
Further properties	voltage-dependent (max. transport rate at 35 mV); uptake stimulated strongly by BSA. Efflux very slow	Efflux fast, DIDS sensitive	fumarate more effective than malate in stimulating ATP-dependent vesicle acidification

a) Martinoia et al., 1985, b) Martinoia et al., 1990, c) Martinoia et al., 1990, d) Rentsch and Martinoia, 1991, e) Marigo et al., 1988, f) Bouyssou et al., 1990, g) Buser-Sutter et al., 1981, h) Nishida et al., 1987, i) White and Smith, 1989.

103

As shown for *Kalanchoe daigremontiana*, uptake of malic acid into barley mesophyll vacuoles was not specific for the natural enantiomer, L-malic acid. D-malic acid, citrate and other di- and tricarboxylic acids acted as competitive inhibitors. Similar results were published by White and Smith (1989) for *Kalanchoe* and by Marigo *et al.*, (1988) for *Catharanthus* cells (Table 1). A dicarboxylic acid worthy of note in this context is MACC (*N*-malonyl-1-aminocyclopropane-1-carboxylic acid) which is synthesised in plant tissues with a high ACC (1-aminocyclopropane-1-carboxylic acid) production (Amrhein, Schneebeck, Skorupka, Tophof, and Stöckigt, 1981). Inhibition experiments with barley suggested that this compound is transferred into the vacuole by the same carrier as malate (Tophof, Martinoia, Kaiser, Hartung, and Amrhein, 1989), whereas a separate carrier has been proposed by Bouzayen, Latche, Alibert, and Pech (1989) in vacuoles isolated from *Catharanthus roseus*. As shown in Table 1, the different malate carriers exhibit different affinities for a variety of mono-, di- and tricarboxylic acids. It is difficult to compare the data from uptake experiments using labelled malate with those where the dissipation of a proton gradient was used to estimate the transport rate of the respective organic acid. Some dicarboxylic acids, such as tartronate or phenylsuccinic acid, may be strong competitive inhibitors but, nevertheless, do not cross the tonoplast (White and Smith, 1989; Martinoia, Vogt, and Amrhein, 1990). On the other hand, it has not been proven that a dicarboxylic acid such as fumarate, which stimulates the generation of a ΔpH across the tonoplast much faster than malate, crosses the tonoplast by the malate carrier.

In *Catharanthus* vacuoles, malate uptake is strongly increased at acidic pH (Marigo *et al.*, 1988; Bouyssou, Canut, and Marigo, 1990). A similar, but weak pH effect has also been observed in barley vacuoles (Rentsch and Martinoia, 1991). The increased uptake at lower pH may indicate that the Hmal⁻ form is recognised by the carrier. Rentsch and Martinoia (1991) determined the concentration-dependent uptake of malate with the incubation medium at a different pH and subsequently calculated the K_m values for the different species of malate, assuming that only a single species of malate was transported across the tonoplast. The results showed that the K_m remained approximately constant for the mal²⁻ form, whereas it changed drastically if the Hmal⁻ was assumed to be the transported form. These results lead to the conclusion that mal²⁻ was the transported form in barley. The situation is less clear for *Catharanthus*, vacuoles, where a strong decrease in the K_m for malate could be observed if the pH of the incubation medium was decreased from 7.5 to 5.5. However, such a result may also be due to the action of an essential amino acid which changes its protonation state in the investigated pH range. Histidine is the only possible candidate and indeed, in both organisms the action of a histidine group has been proposed, since transport is efficiently inhibited by diethylpyrocarbonate (Rentsch and Martinoia, 1991; Dietz, personal communication). Uptake was also inhibited by DIDS and the lysine specific protein modifying agent pyridoxal phosphate (Martinoia *et al.*, 1990). Malate transport could be protected against pyridoxal phosphate if a competitive substrate such as 1,2,3-benzenetricarboxylic acid was present in the pre-incubation medium, indicating that a lysine residue is also present at the substrate binding site. Contradictory results have been published concerning the efflux of malate from the vacuole. While the efflux of malate is very low in intact barley vacuoles (Kaiser, Martinoia, Schröppel-Meier, and Heber, 1989), a fast, DIDS-sensitive efflux was observed in tonoplast vesicles of *Catharanthus* (Bouyssou *et al.*, 1990). For CAM plants, it has been proposed that the uncharged H_2mal form crosses the tonoplast passively during the daytime (Lüttge and Smith, 1984).

The demonstration of a channel which is permeable to monovalent cations, as well as to anions such as chloride and malate (Hedrich, Flügge, and Fernandez, 1986; Coyaud,

Kurkdjian, Kado, and Hedrich, 1987), lead to speculation as to whether these compounds all cross the tonoplast by the same carrier. However, the inhibition of malate uptake by chloride, and *vice versa*, as observed in isolated vacuoles may be due to competition for the energy source, rather than to a common binding site (Martinoia *et al.*, 1990). Impermeable competitive inhibitors of malate uptake, such as tartronate, phenylsuccinic acid or 1,2,3-benzenetricarboxylic acid, do not inhibit chloride uptake (White and Smith, 1989; Martinoia *et al.*, 1990). Additional proof of the distinct nature of the malate and chloride uptake systems is the fact that uptake of malate is strongly inhibited by pyridoxal phosphate, whereas that of chloride is not (Martinoia *et al.*, 1990).

A first step towards identifying the malate carrier was made by Martinoia, Vogt, Rentsch, and Amrhein, (1991) who succeeded in the functional reconstitution of the highly purified malate carrier. Purification was achieved by chromatography on hydroxyapatite followed by affinity chromatography using 5-amino-1,2,3-benzenetricarboxylic acid as a ligand. Although a 30,000 fold purification was achieved, 5 polypeptides were still visible on a silver stained SDS gel (Figure 2).

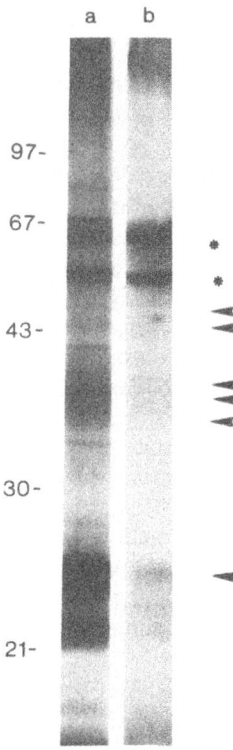

Figure 2 Polypeptide pattern of a hydroxyapatite eluate (A) and an affinity chromatography eluate (B) of tonoplast membranes containing the malate transport activity. The arrows indicate the polypeptides still visible after the affinity chromatography step. The two stained regions stained with an asterisk are due to the buffer system used and do not represent vacuolar polypeptides. For further details see Martinoia *et al.*, (1991).

The carrier, reconstituted in asolectin liposomes, had properties similar to those described previously for the carrier in intact vacuoles (Martinoia *et al.*, 1985). Malate uptake is voltage-dependent with a maximal rate at a membrane potential around +35 mV. This observation may indicate that the transport of malate across the tonoplast is regulated by the membrane potential across this membrane. Different approaches to identifying the polypeptide(s) responsible for malate transport have been taken: i) labelling with pyridoxal phosphate followed by reduction with NaB^3H_4 (Flügge and Heldt, 1977), and ii) 5-(1-hydroxy-4-azidophenylazo)-1,2,3-benzenetri-carboxylic acid, which is a very strong competitive inhibitor of malate transport in barley vacuoles (K_i approximately 5 μM), was iodinated and used as a photoaffinity probe (Rentsch, Vogt and Martinoia, unpublished results). However, the polypeptide(s) involved in malate transport are still unknown. Preliminary experiments, in which different segments of an SDS polyacrylamide gel were tested for malate transport activity, indicate that the carrier has a molecular weight of about 30 kD (Rentsch, Vogt, and Martinoia, unpublished results).

The mechanisms responsible for the control of the accumulation and release of malate *in vivo* have not yet been identified. Reduction of nitrate in the cytosol, which is accompanied by vacuolar efflux (Martinoia, Heck, and Wiemken, 1981) leads to an accumulation of malate in the vacuole (Marigo, Bouyssou, and Belkoura, 1985; Kaiser and Föster, 1989). Enhanced malic acid sythesis during nitrate reduction may be a mechanism which controls the cytosolic pH. The exchange of nitrate and malate was interpreted in terms of an antiport mechanism of these two anions. However, recent results (Blom-Zandstra, Koot, Hattum and Borstlap, 1990; Kaiser and Martinoia, unpublished) suggest that the exchange does not take place by a coupled antiport mechanism but by two distinct carriers. It is interesting that not only the malate content, but also the malate transport activity across the tonoplast, appears to be a function of nitrate content and reduction. Under different growth conditions the malate transport activity could be enhanced three-to four-fold (Martinoia and Vogt, unpublished results). A further aspect of malate compartmentation is the fact that malate is accumulated as the free acid in CAM plants (Lüttge *et al.*, 1981). In contrast, titratable acidity remains low in C3 or C4 plants, even if malate is accumulated at high concentrations (Winter *et al.*, 1982) indicating that malate is present as a K^+ or Na^+ salt in these plants.

Citric acid is present at very high concentrations in CAM plants and in some fruits. It is not widely known that many CAM plants show diurnal fluctuations, not only in malic acid but also in citric acid. This phenomenon has not yet been investigated in detail (Lüttge, 1988). However, vacuolar citrate may play an important role, even at lower concentrations. Therefore, due to its strong tendency to form complexes with divalent cations, citrate may decrease the free concentration of calcium and magnesium within the vacuole and contribute to the vacuolar accumulation of these cations. Uptake of citrate was first investigated in lutoids of *Hevea brasiliensis*, in which it is present at high concentrations (more than 50 mM)(Marin, Cretin, and D'Auzac, 1982). The uptake system has a K_m of about 6 mM and, using pH sensitive dyes, it was suggested that the MgATP-dependent accumulation of citrate occurs by a $H_2citrate^-/H^+$ antiport mechanism. Citrate uptake into vesicles of tomato fruit tonoplasts could be inhibited by tri-and dicarboxylic acids (Oleski, Mahdavi, and Bennet, 1987) and therefore shows similarities to the malate uptake system described by other authors. However, MgATP does not stimulate citrate uptake. Two saturable components were observed, possibly because both citrate^{3-} and Hcitrate $^{2-}$ cross the tonoplast by the same carrier but with different affinities. Citrate uptake into barley mesophyll vacuoles (Rentsch and Martinoia, 1991) exhibits only one saturable component (K_m 0.2 mM) and uptake is stimulated by MgATP. Mg^{2+} inhibited citrate uptake indicating that citrate is not transported as a Mg^{2+} complex. It was suggested that citrate^{3-} is the transported form for tomato and barley. Various di- and

Table 2 Properties of the citrate carrier from different plants.

	Hevea brasiliensis[a]	tomato[b]	barley[c]
K_m	6 mM	125 μM and 1.5 mM	0.2 mM
Inhibition of citrate transport by carboxylic acids		malate = succinate > isocitrate > fumarate >> oxaloacetate	1,2,3 benzenetricarboxylic acid = phenylsuccinate > cis-aconitate = malate > fumarate > malonate = succinate = isocitrate
Inhibitors (concentration), % inhibition		DIDS (10 μM 10%	DIDS (50-200 μM) 85-90% Pyridoxal phosphate (200-500 μM) 60-80% Diethylpyrocarbonate (1mM) 70% PCMBS (1 mM) 98%
	postulated transport mechanism: H_2citrate/H^+ antiport		
Further properties		transported form citrate[3-]	transported form citrate[3-]

a) Marin et al., 1982, b) Oleski et al., 1987, c) Rentsch and Martinoia, 1991.

tricarboxylates inhibit the uptake of malate and citrate to similar extents (Table 2). Such observations suggest that malate and citrate may cross the membrane by the same permease.

Assuming that most of the di- and tricarboxylic acids cross the tonoplast by the same carrier, the question still remains open as to whether monocarboxylic acids, which may be present in high amounts in plant cells (Grob and Matile, 1980; Holländer-Czytko and Amrhein, 1983), have a separate uptake system. The fact that inhibition of malate uptake by monocarboxylates is weak suggests the presence of separate carriers. However, no experimental data are yet available.

REFERENCES

AMRHEIN, N., SCHNEEBECK, D., SKORUPKA, H., TOPHOF, S., and STÖCKIGT, J., 1981. Identification of a major metabolite of the ethylene precursor 1-aminocyclopropane-1-carboxylic acid in higher plants. *Naturwissenschaften*, **68**, 619-620.

BLOM-ZANDSTRA, M., KOOT, H.T.M., HATTUM, J., and BORSTLAP, A.C., 1990. Interactions of uptake of malate and nitrate into isolated vacuoles from lettuce leaves. *Planta*, **183**, 10-16.

BOUYSSOU, H., CANUT, H., and MARIGO, G., 1990. A reversible carrier mediates the transport of malate at the tonoplast of *Catharanthus roseus* ells. *FEBS Letters*, **275**, 73-76.

BOUZAYEN, M., LATCHE, A., ALIBERT, G., and PECH, J.C., 1989. Carrier mediated uptake of 1-(malonylamino)cyclopropane-1-carboxylic acid in vacuoles isolated from *Catharanthus roseus* cells. *Plant Physiology*, **91**, 1317-1322.

BUSER, C., and MATILE, P., 1977. Malic acid in vacuoles isolated from Bryophyllum leaf cells. *Zeitschrift für Pflanzenphysiologie*, **82**, 462-466.

BUSER-SUTTER, C., WIEMKEN, A., and MATILE, P., 1981. A malic acid permease in isolated vacuoles of a crassulacean acid metabolism plant. *Plant Physiology*, **69**, 456-459.

COYAUD, L., KURKDJIAN, A., KADO, R., and HEDRICH, R., 1987. Ion channels and ATP-driven pumps involved in ion transport across the tonoplast of sugar beet vacuoles. *Biochimica et Biophysica Acta*, **902**, 263-268.

FLÜGGE, U.I., and HELDT, H.W., 1977. Specific labelling of a protein involved in phosphate transport of chloroplasts by pyridoxal-5-phosphate. *FEBS Letters*, **82**, 29-33.

GERHARDT, R., STITT, M., and HELDT, H.W., 1987. Subcellular metabolite levels in spinach leaves. *Plant Physiology*, **83**, 399-407.

GROB, K., and MATILE, P., 1980. Compartmentation of ascorbic acid in vacuoles of horseradish root cells. Note on vacuolar peroxidase. *Zeitschrift für Pflanzenphysiologie*, **98**, 235-243.

HEDRICH, R., FLÜGGE, U.I., and FERNANDEZ, J.M., 1986. Patch-clamp studies of ion transport in isolated plant vacuoles. *FEBS Letters*, **204**, 228-232.

HOLLÄNDER-CZYTKO, H., and AMRHEIN, N., 1983. Subcellular compartmentation of shikimic acid and phenylalanine in buckwheat cell suspension cultures grown in the presence of shikimate pathway inhibitors. *Plant Science Letters*, **29**, 89-96.

KAISER, G., MARTINOIA, E., and WIEMKEN, A., 1982. Rapid appearance of photosynthetic products in the vacuoles isolated from barley mesophyll protoplasts by a new fast method. *Zeitschrift für Pflanzenphysiologie*, **107**, 103-113.

KAISER, G., MARTINOIA, E., SCHRÖPPEL-MEIER, G., and HEBER, U., 1989. Active transport of sulfate into the vacuole of plant cells provides halotolerance and can detoxify SO_2. *Journal of Plant Physiology*, **133**, 756-763.

KAISER, W., and FÖRSTER, J., 1989. Low CO_2 prevents nitrate reduction in leaves. *Plant Physiology*, **91**, 970-974.

KÄSTNER, K.H., and SZE, H., 1987. Potential-dependent anion transport in tonoplast vesicles from oat roots. *Plant Physiology*, **83**, 483-489.

LÜTTGE, U., SMITH, J.A.C., MARIGO, G., and OSMOND, C.B., 1981. Energetics of malate accumulation in the vacuoles of *Kalanchoe tubiflora* cells. *FEBS Letters*, **126**, 81-84.

LÜTTGE, U., and SMITH, J.A.C., 1984. Mechanism of passive malic-acid efflux from vacuoles of the CAM plant *Kalanchoe daigremontiana*. *Journal of Membrane Biology*, **81**, 149-158.

LÜTTGE, U., 1988. Day-night changes of citric-acid levels in crassulacean acid metabolism: phenomenon and ecophysiological significant. *Plant Cell and Environment*, **11**, 445-451.

MARIGO, C., COUYSSOU, H., and BELKOURA, M. 1985. Vacuolar efflux of malate and its influence on nitrate accumulation in *Catharanthus roseus* cells. *Plant Science*, **39**, 97-103.

MARIGO, G., BOUYSSOU, H., and LABORIE, D., 1988. Evidence for malate transport into vacuoles isolated from *Catharanthus roseus* cells. *Botanica Acta*, **101**, 187-191.

MARIN, B., CRETIN, H., and D'AUZAC, J., 1982. Energisation of solute transport and accumulation at the tonoplast in *Hevea* latex. *Physiologie Vegetale*, **20**, 333-346.

MARQUARDT-JARCZYK, G., and LÜTTGE, U., 1990. Anion transport at the tonoplast of mesophyll cells of the CAM plant *Kalanchoe daigremontiana*. *Journal of Plant Physiology*, **136**, 129-136.

MARTINOIA, E., HECK. U., and WIEMKEN, A., 1981. Vacuoles as storage compartments of nitrate in barley leaves. *Nature*, **289**, 292-294.

MARTINOIA, E., FLÜGGE, U.I., KAISER, G., HEBER, U., and HELDT, H.W., 1985. Energy-dependent uptake of malate into vacuoles isolated from barley mesophyll photoplasts. *Biochimica et Biophysica Acta*, **806**, 311-319.

MARTINOIA, E., VOGT, E., and AMRHEIN, N., 1990. Transport of malate and chloride into barley mesophyll vacuoles. Different carriers are involved. *FEBS Letters*, **261**, 109-111.

MARTINOIA, E., VOGT, E., RENTSCH, D., and AMRHEIN, N., 1991. Functional reconstitution of the malate carrier of barley mesophyll vacuoles in liposomes. *Biochimica et Biophysica Acta*, **1062**, 271-278.

NISHIDA, K., and TOMINAGA, P., 1987. Energy-dependent uptake of malate into vacuoles isolated from CAM plants *Kalanchoe daigremontiana*. *Journal of Plant Physiology*, **127**, 385-393.

OLESKI, N.M MAHDAVI, P., and BENNETT, A.B., 1987. Transport properties of the tomato fruit tonoplast. II. Citrate transport. *Plant Physiology*, **84**, 997-1000.

POPE, A.J., and LEIGH, R.A., 1987. Some characteristics of anion transport at the tonoplast of oat roots, determined from the effects of anions on pyrophosphate-dependent proton transport. *Planta*, **172**, 91-100.

REA, P.A., and SANDERS, E., 1987. Tonoplast energisation: two H$^+$ pumps, one membrane. *Physiologia Plantarum*, **71**, 131-141.

RENTSCH, D., and MARTINOIA, E., 1991. Citrate transport into barley mesophyll vacuoles - comparison with malate uptake activity. *Planta*, **184**, 532-537.

SCHNABL, H., and KOTTMEIER, C., 1984. Determination of malate levels during the swelling of vacuoles isolated from guard-cell protoplasts. *Planta*, **161**, 27-31.

SZE, H., 1985. H$^+$-translocating ATPases: advances using membrane vesicles. *Annual Review of Plant Physiology*, **36**, 175-208.

TOPHOF, S., MARTINOIA, E., KAISER, G., HARTUNG, W., and AMRHEIN, N., 1989. Compartmentation and transport of 1-aminocyclopropane-1-carboxylic acid and N-malonyl-1-aminocyclopropane-1-carboxylic acid in barley and wheat mesophyll cells and protoplasts. *Plant Physiology*, **75**, 333-339.

WHITE, P.J., and SMITH, J.A.C., 1989. Proton and anion transport at the tonoplast in crassulacean-acid-metabolism plants: specificity of the malate-influx system in *Kalanchoe daigremontiana*. *Planta*, **179**, 265-274.

WINTER, K., USUDA, H., TSUZUKI, M., SCHMITT, M., EDWARDS, G.E., THOMAS, R.J., and EVERT, R.F., 1982. Influence of nitrate aand ammonia on photosynthetic characteristics and leaf anatomy of *Moricandia arvensis*. *Plant Physiology*, **70**, 615-625.

Channel Proteins

Introduction

Ion channels may prove to be the most elusive of membrane proteins. Depending on the size of the conductance pathway they create, fluxes through channels range from large to very large, in comparison with those through other sorts of transport proteins. This works to the advantage of the biophysicist who can observe, through patch-clamping, the conductance of individual channels. Such studies show how the gating of channels is regulated by voltage, by physical stretching of the membrane, by calcium and a growing battery of natural and synthetic effector molecules and channel blockers. In a short space of time we have moved from knowing nothing about the channels in plant membranes to knowing a great deal more about their regulation than we do about the secondary ion and metabolite transporters discussed in the previous section. However, there is a gulf between the biophysical identity of channels and the knowledge of their structure. For all the technical wizardry of modern patch-clamping, it is still a description of transport from the 'outside'. Molecular structures are needed for an 'inside' view.

One plausible approach to finding out something about the polypeptides which comprise channels is from phenotypic complementation of mutants with defective channel activity, eg. Shaker mutants in *Drosophila*, or K^+-transport mutants of yeast. Such mutants are not lethal and can, therefore, be readily exploited. The first contribution in this section was really made possible because of an unusual abundance of a channel protein. The nodulin 26 gene from soybean codes for a channel protein in the peribacteriodal membrane of nodules. This membrane presides over a massive efflux of malate from the host to the bacteriods, the diffusion gradient for which is maintained by high rates of bacteroid respiration. The deduced amino acid sequence of this protein shows an intriguingly high degree of homology with channel proteins of very diverse origin (Verma). This homology counters a suggestion which arose in discussion, that the product of the nodulin 26 gene was a facilitator or ion exchanger rather than a true channel protein. Transformation of the host with antisense genes for nodulin 26 causes ineffective nodulation and peribacteroidal membrane breakdown. Clearly, the successful outcome of the symbiosis depends on the presence of this protein.

The gulf referred to above, between the biophysical and molecular identities of the channel proteins was evident from the vigorous discussions following presentations by Brosnan and Sanders, and Ranjeva *et al*. Many channels can have their conductance blocked by ligands of drugs and toxins; this much is very clear from patch-clamping studies. Confusion became evident in discussion when the nature of these interactions was discussed from a biochemical standpoint. Drugs, such as verapamil, which unquestionably block calcium channels, appear to bind to many membrane proteins. Thus, overall verapamil binding gives an impression that plant cells have many more calcium channels than would be expected from what is known about animal cells. Further uncertainty about the nature of calcium channels was introduced by D.Sanders (University of York, UK) and S.Assman

(Harvard University, USA) who indicated that verapamil has been shown to have dramatic effects on both inwardly and outwardly rectified K^+ currents in *Amaranthus* and that calcium movement can occur via inwardly rectified K^+ channels in guard cells. Therefore, the assumption that the inhibitor of a channel necessarily binds to the channel protein itself needs to be closely examined. Receptor controlled channels are known to exist in animals and show that other proteins can be involved. This is illustrated in the contribution by Brosnan and Sanders which suggests that a receptor for inositol trisphosphate may regulate the opening of calcium channels in the tonoplast in plant cells.

Soybean Nodulin-26: A Channel Protein Conserved from Bacteria to Mammals

D.P.S. Verma

Department of Molecular Genetics and Biotechnology Centre
The Ohio State University
Columbus, OH 43210, USA

Abstract

Nodulin-26 is an intrinsic membrane protein of the peribacteroid membrane in soybean root nodules. This protein contains six transmembrane domains and forms an ion channel translocating specific molecules to the bacteroids. We have determined the topology of this protein by co-translational processing and protease protection experiments. Our results suggest that the carboxy and amino termini of this protein face the cytoplasm while the glycosylation site at amino acid position 150 is located on the surface facing the bacteroids. Two transmembrane domains are sufficient for the integration of nodulin-26 into the membrane as revealed by deletion analysis. Nodulin-26 is phosphorylated by a kinase located in the peribacteroid membrane. The phosphorylation mechanism may allow nodulin-26 to be an active channel.

Keywords

Peribacteroid membrane, nitrogen fixation, protein phosphorylation, ion transport.

A family of structurally-related intrinsic membrane proteins has recently been identified from diverse organisms. These proteins form channels in different membranes and appear to be synthesised in specialised tissues and organs for transport of specific solutes. Ten members of this conserved family of proteins have been characterised. These include: GlpF of *Escherichia coli*, forming a pore type channel in the inner membrane (Maramatsu and Mizuno, 1989; Sweet, Gandor, Voegele, Wittekindt, Beuerle, Truniger, Lin, and Boos, 1990); MIP, major intrinsic protein of bovine (Gorin, Yanc, Cline, Revel, and Horowitz, 1984) and rat (Kent and Shiels, 1990) lens fibre membrane, nodulin-26 of soybean, a plant gene-encoded protein of the peribacteroid membrane in root nodules

Transport and Receptor Proteins of Plant Membranes
Edited by D.T. Cooke and D.T. Clarkson, Plenum Press, New York, 1992

113

(Fortin, Morrison, and Verma, 1987); TIP, a tonoplast membrane protein synthesized in protein bodies of bean (Johnson, Hofte, and Chrispeels, 1990); BIB, a neurogenic protein of *Drosophila* (Rao, Jan and Jan, 1990); Tob Rb7, a root tip-specific protein from tobacco and *Arabidopsis* (Yamamoto, Cheng, and Conkling, 1990); and JM7a, a protein encoded by a turgor-regulated gene in Pea (Guerrero, Jones, and Mullet, 1990). In addition, homologs of these proteins have been isolated from yeast (Van Aelst, Hohmann, Zimmermann, Jans, and Thevelein, 1991) and filamentous fungi (*Streptomyces coelicolor*; Smith and Chater, 1988).

With the exception of BIB, these proteins are similar in size (250-300 amino acids) and exhibit similar hydropathy profiles containing six potential membrane-spanning domains. These proteins are also extremely hydrophobic, as shown by the experimental data on MIP (Wong, Robertson, and Horowitz, 1978) and nodulin-26 (Miao, Hong, and Verma, 1991) and are soluble only in the lipid phase of Triton X-114. The GlpF and nodulin-26 have similar isoelectric points (9.6) while the isoelectric point of TIP is similar to that of MIP (Verma, Miao, Joshi, Cheon, and Delanney, 1990). These proteins appear to function as homotetramers. The subcellular location of these proteins varies in each organism; for example, MIP is located in the plasma membrane of the lens fibre cells and TIP in tonoplast membrane (Johnson *et al.*, 1990).

Nodulin-26 was identified from the sequence analysis of a cDNA clone abundantly expressed in soybean root nodules (Fortin *et al.*, 1987) and was shown to be an integral protein of the membrane enclosing the bacteria (peribacteroid membrane, PBM) in the host cell. This subcellular compartment is formed *de novo* following infection of the plant by *Rhizobium*. The PBM forms a primary interface between the bacteria and the plant cell (Verma and Fortin, 1989). Although derived from the host plasma membrane during endocytosis of bacteria, the properties of the PBM appear to change and it behaves as a mosaic membrane. Many properties of this subcellular compartment suggest that it is a vacuole type as revealed by the presence of some vacuolar marker proteins such as protease inhibitors (see Manen, Simon, Van Slooten, Osteras, Fruitiger, and Hughes, 1990), while many plasma membrane proteins also continue to exist in the PBM (Fortin *et al.*, 1987). A H^+-ATPase was localised in this membrane and was found to be ammonia-stimulated (Blumwald, Fortin, Rea, Verma, and Poole, 1986). In addition, physical properties (e.g. thickness and phosphotungstic acid stainability) are similar to those of the plasma membrane.

We have determined the mechanism of integration and topology of nodulin-26 by *in vitro* translation of its mRNA in the absence or presence of microsomal membranes followed by the treatment of membrane vesicles by trypsin (Figure 1). The data suggest (Miao *et al.*, 1991) that nodulin-26 is co-translationally inserted into the membrane without cleavage of any signal sequence and is glycosylated in the endoplasmic reticulum. Both carboxy and amino termini face the cytoplasm, as they are sensitive to tryptic digestion. The glycosylation site exists on the surface facing the bacteria and is glycosylated as revealed by binding with Con-A. This protein is phosphorylated at its carboxy end where three potential phosphorylation sites (serine) exist. We have shown that the kinase phosphorylating the residues co-exists in the PBM. The activity of this enzyme is Ca^{2+}-dependent but calmodulin-independent. Similarly, MIP and TIP are also phosphorylated but GlpF is not, suggesting that these channels may have become further specialised to meet the specific metabolic needs of the tissue in which they are found to be synthesised. With the exception of GlpF, the function of no other member of this gene family is known. By definition (Legocki and Verma, 1980), nodulin-26 is expressed only in root nodules and is not detectable in any other part of the plant. However, using specific primers corresponding to the conserved regions of nodulin-26, a low-abundant transcription has been detected in the shoot, flower and root meristems of soybean (Miao, G.-H. and D.P.S. Verma, unpublished data). This sequence may belong to other members of this gene family as many homologs appear to exist in the soybean genome. We have also isolated a homolog of nodulin-26 from *Vigna aconitifolia* (mothbean).

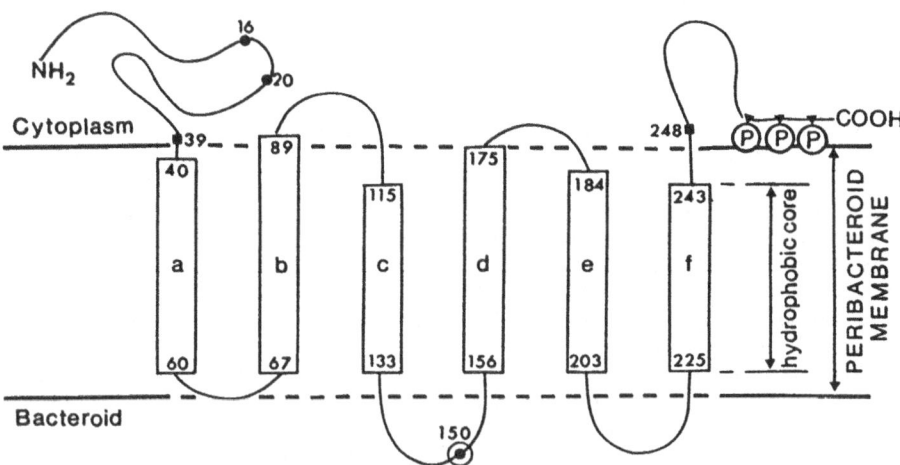

Figure 1 Topology of soybean nodulin-26 (See Miao *et al.*, 1991): a-f, six membrane-spanning domains, determined by computer analysis. (▼), Phosphorylation sites; (●), Potential glycosylation sites that are not glycosylated; (○), Glycosylation site facing the bacteroid is in fact glycosylated as shown by Con-A binding; (■), Trypsin target sites closest to the first and last transmembrane domains; (P), Phosphate residues interacting with the membrane.

Determining the function of nodulin-26 in the PBM is of paramount importance in order to understand the flow of various metabolites from the plant cell to the bacteroids. If this channel acts like GlpF, it may allow translocation of various ions from the cytoplasm to the peribacteroid space (PBS) from which an active uptake by bacteroids can then follow. The GlpF is known to transport many compounds with a size limit of 0.4 nM (Heller, Lin, and Wilson, 1980). Among the most abundant solutes in root nodules, dicarboxylic acids (malate and succinate) are actively transported across the PBM. Preliminary studies suggest that this transport is not inhibited by antibodies against nodulin-26. Myo-ionositol also exists in significant quantities in the host cytosol (Streeter, 1990), but it is not taken up easily by the peribacteroid units (K. Chapman and D.P.S. Verma, unpublished data).

Isolation of the gene encoding nodulin-26 and transfer of its promoter following a reporter gene (GUS) fusion showed that soybean nodulin-26 is expressed in incipient root primordia and root tips (G.-H. Miao and D.P.S. Verma, manuscript in preparation). This is consistent with the presence of another member of this gene family, Tob RB7, in tobacco and *Arabidopsis* root tips (Yamamoto *et al.*, 1990). Further studies of this family of transport proteins may not only shed light on the evolution of these genes, but will also show how each molecule has adapted to transport specific compounds and is regulated. The co-existence of a specific kinase phosphorylating some members of this family provides further regulatory control of this channel. In addition, promoter analysis of nodulin-26 may help decipher the induction mechanism and the role of metabolites, if any, in the induction of this gene.

Acknowledgement

This work was supported by a research grant from the NSF (DCB-8819399).

REFERENCES

BLUMWALD, E., FORTIN, M.G., REA, VERMA, D.P.S., and POOLE, R.J., 1985. Presence of host-plasma membrane type H^+-ATPase in the membrane envelope enclosing the bacteroids in soybean root nodules. *Plant Physiology*, **78**, 665-672.

FORTIN, M.G., MORRISON, N.A., and VERMA, D.P.S., 1987. Nodulin-26, a peribacteroid membrane nodulin, is expressed independently of the development of the peribacteroid compartment. *Nucleic Acids Research*, **15**, 813-824.

GORIN, M.B., YANCEY, S.B., CLINE, J., REVEL, J.-P. and HORWITZ, J., 1984. The major intrinsic protein (MIP) of the bovine lens fibre membrane: characterisation and structure based on cDNA cloning. *Cell*, **39**, 49-59.

GUERRERO, F.D., JONES, J.T., and MULLET, J.E., 1990. Turgor-responsive gene transcription and RNA levels increase rapidly when pea shoots are wilted. Sequence and expression of three inducible genes. *Plant Molecular Biology*, **15**, 11-26.

HELLER, K.B., LIN, E.C.C., and WILSON, T.H., 1980. Substrate specificity and transport properties of the glycerol facilitator of *Escherichia coli*. *Journal of Bacteriology*, **144**, 274-278.

JOHNSON, K.D., HOFTE, H., and CHRISPEELS, M.J., 1990. An intrinsic tonoplast protein of protein storage vacuoles in seeds is structurally related to a bacterial solute transporter (GlpF). *The Plant Cell*, **2**, 525-532.

KENT, N.A., and SHIELS, A., 1990. Nucleotide and derived amino acid sequence of the major intrinsic protein of rat eye lens. *Nucleic Acids Research*, **18**, 4256.

LEGOCKI, R.P., and VERMA, D.P.S., 1980. Identification of "nodule-specific" host proteins (nodulins) involved in the development of *Rhizobium*-legume symbiosis. *Cell*, **20**, 153-163.

MANEN, J.-F., SIMON, P., VAN SLOOTEN, J.-C., OSTERAS, M., FRUITIGER, S., and HUGHES, G.J., 1991. A nodulin specifically expressed in senescent nodules of winged bean is a protease inhibitor. *The Plant Cell*, **3**, 259-270.

MIAO, G.-H., HONG, Z., and VERMA, D.P.S., 1991. Insertion, topology and phosphorylation of soybean nodulin-26 in the peribacteroid membrane (submitted).

MURAMATSU, S., and MIZUNO, T., 1989. Nucleotide sequence of the region encompassing the *glp*KF operon and its upstream region containing a bend DNA sequence of *Escherichia coli*. *Nucleic Acids Research*, **17**, 4378.

RAO, Y., JAN, L.Y., and JAN, Y.N., 1990. Similarity of the product of the *Drosophila* neurogenic gene big brain to transmembrane channel proteins. *Nature*, **345**, 163-167.

SMITH, C.P., and CHATER, K.F., 1988. Structure and regulation of controlling sequences for the *Streptomyces coelicolor* glycerol operon. *Journal of Molecular Biology*, **204**, 569-580.

STREETER, J.G., 1987. Carbohydrate, organic acid and amino acid composition of bacteroids and cytosol from soybean nodules. *Plant Physiology*, **85**, 768-773.

SWEET, G., GANDOR, C., VOEGELE, R., WITTEKINDT, N., BEUERLE, J., TRUNIGER, V., LIN, E.C.C., and BOOS, W., 1990. Glycerol facilitator of *Escherichia coli*: cloning of glpF and identification of the glpF product. *Journal of Bacteriology*, **172**, 424-430.

VAN AELST, L., HOHMANN, S., ZIMMERMANN, F.K., JANS, A.W.H., and THEVELEIN, J., 1991. A yeast homologue of the bovine lens fibre MIP gene family complements the growth defect of a *Saccharomyces cerevisiae* mutant on fermentable sugars but not its defect in glucose-induced RAS-mediated cAMP signalling. *The EMBO Journal*, **10**, 2095-2104.

VERMA, D.P.S., and FORTIN, M.G., 1989. Nodule development and formation of the endosymbiotic compartment. In *The Molecular Biology of Nuclear Genes*. Eds J. Schell and I.K. Vasil. Academic Press Inc., New York. pp 329-353.

VERMA, D.P.S., MIAO, G.-H., JOSHI, C.P., CHEON, C.-I., and DELAUNEY, A., 1990. Internalisation of *Rhizobium* by plant cells: targeting and role of peribacteroid membrane nodulins. In *Plant Molecular Biology 1990*. Eds R. Herrman and B. Larkins. Plenum Press. pp 121-130 (in press)

WONG, M.M., ROBERTSON, N.P., and HORWITZ, J., 1987. Heat induced aggregation of the sodium dodecyl sulfate solubilised main intrinsic polypeptide isolated from bovine lens plasma membrane. *Biochemical and Biophysical Research Communications*, **84**, 158-165.

YAMAMOTO, Y.T., CHENG, C.-L. and CONKLING, M.A., 1990. Root-specific genes from tobacco and *Arabidopsis* homologous to an evolutionarily conserved gene family of membrane channel proteins. *Nucleic Acids Research*, **18**, 7449.

The Application of Planar Lipid Bilayers to the Study of Plant Ion Channels

Philip J. White[1] and Mark Tester[2]

1.   Department of Botany
University of Cambridge
Downing Street
Cambridge, CB2 3EA, UK

2.   Department of Botany
University of Adelaide
GPO Box 498
Adelaide, SA5001, Australia

Abstract

In this article we describe the basic equipment and techniques required to study plant ion channels in planar lipid bilayers (PLB). We discuss the advantages and drawbacks of this technique compared to other electrophysiological techniques. We review previous PLB studies of plant ion channels and discuss our recent investigation of K^+ channels at the plasma membrane of rye (*Secale cereale*) roots. We conclude with an overview of the possible future applications of this technique.

Keywords

Electrophysiology, ion channels, planar lipid bilayer, plasma membrane, plants.

1.   Introduction

1.1   Ion channels and single channel recording

Ion channels are integral membrane proteins which, when open, allow the flux of ions down their electrochemical gradient at rates in the order of 10^7 ions s^{-1} (Hille, 1984). This generates a measurable electrical current, even through a single channel and provides the basis for electrophysiological studies of excitable membranes. Ion channels are involved in

Transport and Receptor Proteins of Plant Membranes
Edited by D.T. Cooke and D.T. Clarkson, Plenum Press, New York, 1992

119

many physiological processes including nutrient transport and compartmentation, plant action potential generation and subsequent repolarisation, cellular signalling and the regulation of volume, turgor and osmolarity. Thus, an understanding of the biophysical characteristics and cellular control of ion channels will be required to understand fully these processes. This can be effected most elegantly by the study of single ion channels.

The development of methods for studying single ion channels is one of the most important and powerful recent advances in electrophysiology. There are two methods currently available. The first method is membrane-patch voltage-clamping (patch-clamping), which involves the electrical analysis of small patches of membrane sealed on to a micropipette. This technique has been reviewed by Sakmann and Neher, (1983) and, as applied to plants, by Hedrich and Schroeder (1989) and Tester (1990). The second method is the reconstitution of either membrane vesicles or isolated channel proteins into artificial planar lipid bilayers (PLB), (Miller, 1986).

In the study of plant ion channels, patch-clamping has been extensively employed (Hedrich and Schroeder, 1989; Tester, 1990) but, until recently, there have been few studies using PLB (see Section 4). However, the incorporation of ion channels into PLB has particular advantages over patch-clamping both in the study of membranes which are inaccessible to patch-clamp electrodes and for the analysis of purified and reconstituted proteins. These aspects have been successfully exploited in the study of a number of ion channels from animal systems, which are routinely incorporated into PLB (Coronado, 1986; Miller, 1986). In general, ion channels incorporated into PLB have similar properties to those observed in native membranes (see articles in Miller, 1986) and studies of the conductance, permeability, kinetics and pharmacology of ion channels in PLB have been readily extrapolated to both macroscopic electrical observations and physiological processes. This is remarkable given the reduced, artificial nature of the PLB system. However, some cases are known where channels do behave abnormally in PLB. In this article we will describe the basic PLB technique, its advantages and drawbacks with respect to other electrophysiological techniques and how the PLB has been, and might be, applied to the study of plant ion channels. We will illustrate these points with reference to studies which are currently underway in our laboratories.

2. The Planar Lipid Bilayer Technique

2.1 Mechanical and electrical hardware

Planar bilayers may be formed by either of two contrasting methods: painting membranes or folding membranes (reviewed by Alvarez, 1986; Coronado, 1986). In our laboratories we paint bilayers by applying a small amount of lipid dispersed in decane to an aperture drilled into a styrene co-polymer cup which separates two aqueous salt solutions, one contained within the cup (*cis*) and the other within an external perspex chamber (*trans*), (Figure 1). As the lipid solution drains to the border of the aperture, a film is formed in the central region. This film gets thinner and finally becomes a bilayer. In the contrasting 'folding' technique, bilayers are formed by the apposition of two lipid monolayers, which affords the possible advantage of obtaining solvent-free, or at least solvent-reduced, bilayers (Montal and Mueller, 1972; Schindler and Rosenbusch, 1978). But, it is generally easier to fuse membrane vesicles to solvent-containing bilayers.

The electrical potential across the bilayer (termed the membrane potential) is controlled in a process known as voltage-clamping. It is accomplished by an electrical circuit (a feedback amplifier) that measures the voltage across the membrane and constantly adjusts it to match the command voltage that the experimenter wishes to impose. To do this, the feedback amplifier generates a current equal in magnitude but opposite in sign to the current that flows through the membrane. Thus, the feedback amplifier serves two functions; it controls the voltage across the membrane and it monitors the current that flows through the membrane. The amplifier is connected to the bilayer chambers by calomel electrodes and 3 M KCl / 1% agar salt bridges. It is conventional in bilayer studies to record membrane potentials *cis* with respect to *trans*, which is held at ground.

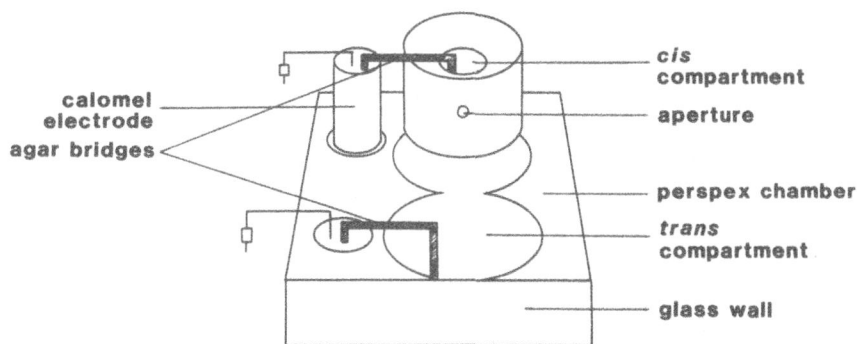

Figure 1 Membrane chamber design for painted planar lipid bilayers.

2.2 The incorporation of channels into the bilayer

In order to study ion channels they must be incorporated into the PLB. By convention, channel-forming material is added only to the *cis* chamber. This ensures a remarkable sidedness of fused channels, either *cis*-cytoplasmic or *cis*-extracellular, but seldom both together (Coronado, 1986). In our research we generally fuse either native membrane vesicles or partially-purified channels reconstituted into liposomes with the PLB. Due to the sensitivity of the system to contaminant channel activity, it is always important to correlate the appearance of channel activity with the addition of the channel-forming material of interest.

Although the exact mechanism of membrane fusion is still unknown, the conditions under which membrane vesicles or liposomes can be induced to fuse to planar lipid bilayers have been determined experimentally (Miller, Arvan, Telford, and Racker, 1976; Cohen, Zimmerberg, and Finkelstein, 1980; Cohen, 1986; Hanke, 1986). Vesicle fusion requires a trans-bilayer osmotic gradient, the side to which the vesicles are added, *cis*, being hyperosmotic (greater in concentration) to the *trans* side (Miller *et al.*, 1976; Cohen *et al.*, 1980; Woodbury and Hall, 1988a; Cohen, Niles, and Akabas, 1989). In addition, fusion requires that phospholipid vesicles are osmotically swollen. Vesicle swelling may be achieved by the presence of active ion channels, or alternatively by utilising membrane-permeant osmoticants such as urea or glycerol (Cohen, 1986). Once consequence of this is that membrane vesicles which contain active ion channels are more likely to fuse to PLB than those which do not (Woodbury and Hall, 1988b; Niles, Cohen, and Finkelstein, 1989). Vesicles fuse to the PLB in a defined manner such that the inside of the vesicle becomes exposed to the *trans* chamber (Figure 2), (Cohen, 1986).

Fusion events can be controlled by manipulating (i) the vesicle concentration in the aqueous phase, (ii) the area of the planar lipid bilayer, (iii) the amount of hyper-osmolarity of the *cis* chamber or vesicles, (iv) the amount of organic solvent in the bilayer, (v) the amount of negatively charged phospholipid in the bilayer, together with the calcium concentration of the aqueous solution, and (vi) the amount of phosphatidylethanolamine in the bilayer (Hanke, 1986).

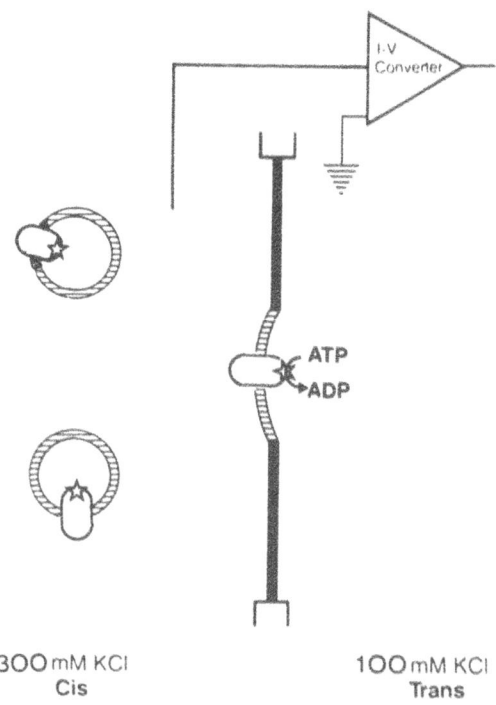

Figure 2 Schematic illustration of the incorporation of plasma membrane vesicles of a single orientation into PE bilayers in stirring in the presence of osmotic gradient. The cartoon illustrates the sidedness of incorporation, the cytoplasmic face of the plasma membrane, represented by the hydrolytic site of the ATPase (★), becoming exposed to the *trans* chamber, which is grounded.

In our standard system, we incorporate vesicles by stirring in the presence of an osmotic gradient (White and Tester, 1991). It has been observed that the solution in the *cis* chamber, to which the vesicles are added, must be agitated for the vesicles to fuse (Cohen, 1986). It is our experience that when stirring is stopped, further channels rarely incorporate into the bilayer. Thus, in practice, fusion may also be controlled by modulating the rate of stirring and stopping stirring when channel activity is detected. In addition, we prevent further channel incorporation by perfusing out the chamber to which the vesicles were added immediately when channel activity is detected.

3. The Advantages and Drawbacks of the Planar Lipid Bilayer Compared to Patch-clamping and other Electrophysiological Techniques

The PLB is a technique for studying single ion channels. Thus, it is complementary to conventional impaling-electrode voltage-clamping and patch-clamping in the whole-cell mode (Miller, 1986; Tester, 1990). Measurements on the ensemble activity of a population of ion channels in any given cellular membrane, such as is obtained by impaling electrophysiology or cell-attached patch-clamping, are usefully applied to the study of whole membrane conductance properties and their modulation by effectors, particulary on a temporal basis. However single channel studies are more appropriate to determine the conductance and subconductance states of individual channels, their mean open lifetimes and open/closed state kinetic relationships and to determine unambiguously the kinetics of interaction of channels with effectors.

Planar lipid bilayers are most advantageous in the study of single channels in membranes not easily accessible to other means of recording, such as the endoplasmic reticulum. The PLB is also a useful technique to study channels in the plasma membrane, as vesicles are prepared without cell wall-digesting enzymes, unlike most protoplasts for patch-clamping; and for studying previously characterised channels in well-defined membrane phospholipids and electrolyte solutions over lengthy time periods (several hours). The PLB is also a useful assay technique *per se*. In general, the purification of ion channel proteins requires a biochemical assay for their activity. Since channels only have a measurable ion transport function when integrated into a membrane between two aqueous compartments, any biochemical assay of channel activity during purification procedures much reconstitute channels into either planar lipid bilayers or liposomes. For the electrical characterisation of isolated ion channels, reconstitution into planar lipid bilayers is probably simpler than patch-clamping giant liposomes (cf. Hedrich, Stoeckel, and Takeda, 1990).

The disadvantages of the PLB technique for the study of ion channels are: (i) The ambiguous origin of ion channels when partially purified membrane and/or protein fractions are used. This is especially pertinent when incorporating vesicles from membrane fractions. A channel in a contaminant membrane may be more abundant or the contaminant membrane itself more fusagenic. (ii) Difficulties in the physiological interpretation of results, firstly since the preparative methods used to isolate ion channels may result in the loss of cellular regulatory mechanisms or alter the annular lipids surrounding the channel and secondly, since the vectorial organisation of the channels may be obscured in vesicular preparations containing channels in diverse orientations. (iii) Difficulties in quantifying ion channel abundance; not only because ion channel-containing vesicles may not fuse with the PLB in amounts which reflect their presence in the preparation, bilayer fusion being influenced both by membrane composition and ion channel activity (see Section 2.2), but also because the fusion of vesicles not containing ion channels cannot be detected. Thus, the insertion of native membrane vesicles into PLB is not a valid method to screen the channel types in a given vesicle population.

These disadvantages can be addressed with varying levels of satisfaction and ease. The best way to circumvent the problem of contaminating protein is to correlate the appearance of particular channels with the addition of preparation enriched in the protein of interest; a suitable "control" is the co-localisation of channel activity with membrane marker-enzymes, immunochemical or specific ligand-binding properties or protein enrichment. For example, to characterise the channels in the endoplasmic reticulum (ER), we assay channel-forming activity along a sucrose gradient and correlate it with the ER marker-enzyme antimycin A-insensitive NADH cytochrome c reductase, both in the presence and absence of Mg^{2+}. In the presence of Mg^{2+}, the ER retains its ribosomes and has a high buoyant density, whereas in the absence of Mg^{2+}, the ER is denuded of ribosomes and has a low buoyant density. Thus, both channel-forming and marker-enzyme activities of the ER will exhibit different buoyant densities in the presence and absence of Mg^{2+}. This Mg^{2+}-shift has been used often to determine whether enzymatic activities are associated with the ER (Lord, 1983) and in using the PLB as an assay for our enzymic (channel) activity, we employ the same technique.

The loss of physiological reality is compensated for by the simplicity of the system, and thus the ability to study directly the effects of putative regulators and lipid composition on channel activity. Problems associated with the orientation of ion channels in the PLB can either be avoided using preparative techniques yielding vesicles of a single orientation (for example plasma membrane prepared by aqueous polymer two-phase partitioning, see Section 4.2) or resolved by comparison with *in vivo* electrical observations, using compounds which interact with only one side of the channel.

Recently, a technique has been devised to circumvent the problems of both the selective incorporation of membrane vesicles with active channels and the detection of vesicle fusion (Woodbury and Miller, 1990). In this technique membrane vesicles are loaded with ergosterol then treated with the polyene antibiotic nystatin. The association of nystatin with ergosterol forms low-conductance, weakly anion-selective channels. Since the nystatin/ergosterol channels are present in all vesicles, this ensure that all vesicles in the preparation have similar probabilities of fusion (Niles and Cohen, 1987). Furthermore, the fusion of any vesicle containing the nystatin/ergosterol channel to the PLB results in a conductance spike which can be counted. Fortunately, upon fusion of a vesicle to the PLB, the nystatin/ergosterol channels inactivate rapidly since the ergosterol diffuses away from the nystatin into the sterol-free PLB, leaving only native ion channels active. Thus, it is now possible to determine the absolute density of active ion channels in a bulk vesicle preparation and it is easier to interpret negative experiments: namely whether the inability to detect channels is due to poor vesicle fusion or the absence of active ion channels (Woodbury and Miller, 1990).

4.   Application of the PLB to the Study of Plant Ion Channels

4.1   Literature review

Plant ion channels have rarely been studied by incorporation into PLB. However, there are examples of the successful incorporation of both isolated membrane fractions and purified proteins of plant, algal or fungal origin into bilayers. In this brief review we have concorded with the bilayer convention and have defined the side to which biological material was added as the *cis* side and referenced the membrane potential *cis* with respect to *trans*.

The first membrane vesicles incorporated into PLB technique were from the outer membrane of the chloroplast envelope. A large conductance channel was incorporated into L-α-phosphatidylcholine bilayers. This porin-type channel was voltage-dependent, closing when the membrane potential was increased from the reversal potential and had a single channel conductance, which was linearly related to KCl concentration, of 720 pS in symmetrical 100 mM KCl (Flügge and Benz, 1984).

An equally large-conductance channel, the voltage-dependent anion channel (VDAC) of corn root mitochondria, has also been characterised following its incorporation into planar azolectin bilayers (Smack and Colombini, 1985; Colombini, 1987). The properties of this channel were strikingly similar to VDACs of animal, protist and fungal mitochondria. It had a unitary conductance of 3.9 ± 0.3 nS in symmetrical 1 M KCl, a weak anion selectivity ($P_{Cl}:P_K = 1.7$) and an open probability which was reduced when the membrane potential was increased from the reversal potential in either the positive or negative direction.

Two types of channels from thylakoids of *Spinacea oleracea* have been studied in planar lipid bilayers. These channels were observed when vesicles enriched in stromal lamellae were incorporated into synthetic 1-palmitoyl-2-oleoyl phosphatidylethanolamine bilayers (Tester and Blatt, 1989). The larger channel had a unitary conductance of 93 pS in asymmetrical 500:300 mM KCl, exhibited voltage-dependence, was selective for K^+ over Cl^- and was inhibited by TEA^+. The smaller channel had a unitary conductance of 14 pS in asymmetrical 500:300 mM KCl. Neither channel appeared to be similar to the Cl^- channels observed by patch-clamping swollen thylakoids from the unusual giant chloroplasts of *Pereromia metallica*, which had a unitary conductance of 80 to 100 pS in symmetrical 100 mM KCl and were voltage-dependent (Schonknecht, Hedrich, Junge, and Raschke, 1988). However, both K^+ and Cl^- channels may be expected to co-exist in this membrane.

In a preliminary investigation of anion channels in membranes from *Beta vulgaris*, a low conductance (unitary conductance 20 pS in symmetrical 750 mM KCl) chloride channel was incorporated into PLB from plasma membrane fractions and weakly selective ($P_{Cl}:P_K = 2.3$) channel with a unitary conductance of 38 pS in 150:750 mM KCl was incorporated from tonoplast fractions (Tester, 1988). The latter channel can be compared with reports of a rather non-selective anion channel at the tonoplast of a variety of plant species observed using patch-clamp techniques (Hedrich, Barbier-Brygoo, Felle, Flügge, Lüttge, Maathuis, Marx, Prins, Raschke, Schnabl, Schroeder, Struve, Taiz, and Ziegler, 1988). In contrast, when vesicles enriched in endoplasmic reticulum from *Beta vulgaris* were incorporated into PLB, a K^+ channel approximating 50 pS in asymmetrical 250:50 mM KCl and a Cl^- channel of 10 pS in symmetrical 100 mM choline chloride were observed (Rousseau and Briskin, 1989).

Although there are as yet no reports of cation channels from the higher plant tonoplast being incorporated into a PLB, a large (435 pS in symmetrical 300 mM KCl) cation channel has been studied in planar azolectin bilayers, following incorporation of right-side-out tonoplast vesicles derived from spheroplasts of *Saccharomyces cerevisiae* (Wada, Ohsumi, Tanifuji, Kasai, and Anraku, 1987; Tanifuji, Sato, Wada, Anraku, and Kasai, 1988). The selectivity of this channel was Cs > K > Na > Li (Wada *et al.*, 1987). The open probability (P_o) of this channel was voltage dependent and channel opening required Ca^{2+}(1 mM) on the *cis* side. Channel opening could be modelled as having two gating mechanisms, which must both be open for the channel to pass current (Tanifuji *et al.*, 1988): a slow responding gate (closing at negative voltages, and requiring Ca^{2+} for opening) and a fast responding gate (closing at positive voltages, and locked open by DIDS). Although recent investigations by patch-clamp techniques also show the existence of Ca^{2+}-dependent channels in sugar-beet tonoplast (Hedrich and Neher, 1987), the gating and conductance characteristics of these channels differ from the yeast channels.

Two partially purified proteins of contrasting origins have been incorporated into planar lipid bilayers: a Ca^{2+} channel from giant algae (Aleksandrov, Berestovsky, Volkova, Vostrikov, Zherelova, Kravchik, and Lunevsky, 1976; Lunevsky, Aleksandrov, Berestovsky, Volkova, Vostrikov, and Zherelova, 1977; Grishchenko, Aleksandrov, and Berestovsky, 1984) and a plasma membrane, verapamil-binding Ca^{2+} channel from coleoptiles of *Zea mays* (Tester and Harvey, 1989). The algal Ca^{2+} channel was studied in planar azolectin bilayers. This channel had a unitary conductance of approximately 200 pS in symmetrical 100 mM KCl and exhibited a selectivity towards divalent cations (Ba > Sr > Ca > Mg) and monovalent cations (Rb > K > Cs > Na > Li). When affects of these ions on gating were also taken into consideration, the channel current could account for the non-chloride inward currents during the plasma membrane action potential in *Nitellopsis obtusa* (Lunevsky, Zherelova, Vostrikov, and Berestovsky, 1983; Grishchenko *et al.*, 1984; Berestovsky, Zherelova, and Kataev, 1987). Chloride did not permeate this channel. The verapamil-binding Ca^{2+} channel was cation selective (Ba > Ca > K) and exhibited a unitary conductance of 12 pS in symmetrical 50 mM $BaCl_2$ (Tester and Harvey, 1989).

4.2   Potassium channels of rye root plasma membrane

At present we are characterising the K^+ channels present in the plasma membrane of rye roots. The electrophysiological characteristics of the root plasma membrane are difficult to study by conventional impaling-electrode voltage-clamping, due to the small dimensions of root cells and their high cell-to-cell, plasmadesmatal conductivity (Grabov, 1990). In addition, the plasma membrane itself is difficult to study by patch-clamping techniques, due to the low frequency of successful sealing of the electrode to protoplasts. Thus, the results presented here complement the sparse literature on the ion channels at the plasma membrane, particularly of root cells.

Rye root plasma membrane vesicles, prepared by aqueous-polymer two-phase partitioning, have been incorporated into planar 1-palmitoyl-2-oleoyl phosphatidylethanolamine bilayers (White and Tester, 1991). Since the plasma membrane vesicles were predominantly right-side-out, and vesicles fuse to bilayers in a predetermined manner (see Section 2.2), plasma membrane channels would be incorporated into the bilayer with their cytoplasmic side facing the *trans* chamber (Figure 2). Thus, since channel orientation can be deduced, these experiments can be compared with patch-clamp or conventional impaling electrophysiological studies.

Five distinct K^+-selective channels were commonly observed in asymmetrical (*cis:trans*) 280:100 mM KCl, which had unitary conductances (determined between +30 mV and -30 mV) of 500, 194, 49, 21 and 10 pS (White and Tester, 1991). The frequency of incorporation of these channels into bilayers was 48, 21, 50, 10 and 9% respectively (data from 159 bilayers). Although these channels have been assumed to originate from the plasma membrane, there is a possibility that some channels, especially those observed in low frequencies, may originate from other, contaminating membranes. However, preliminary evidence suggests that both the 500 pS and the 49 pS channels co-localise with plasma membrane, vanadate-sensitive ATPase activities upon sucrose density gradient centrifugation.

The 500 pS channel was selective for both monovalent and divalent cations over anions (Table 1). Its conductivity and permeability sequences were similar (Table 1), (White 1992a). The conductivity sequence was Rb ≥ K > Cs > Na > Ba > Sr > Li > Ca > Mg > Mn. However, although monovalent cation permeabilities followed the conductivity sequence, all divalent cations exhibited similar permeabilities. The channel was strongly voltage-dependent (Figure 3), (White 1992b). There was increase gating of the channel as membrane potential was increased, which was most marked at negative voltages. Also, the channel was deactivated at increasingly shorter intervals following a voltage step from zero as membrane potential was increased. Channel deactivation did not result from loss of any regulatory factors nor from protein modification, since channel activity was restored immediately upon depolarisation to zero. Thus, the 500 pS channel would open upon depolarisation *in vivo* and facilitate an outwardly-rectified K^+ current, but an inward Ca^{2+} current.

It is interesting to note that several properties of the 500 pS channel have been reported for putative plasma membrane Ca^{2+} channels. Its conductivity (although higher) and selectivity are comparable with the Ca^{2+} channels extracted from giant algae and studied in

Table 1 The selectivity of the 500 pS cation channel from the plasma membrane of rye roots expressed relative to K^+. The permeability of chloride relative to K^+ was estimated from the zero-current potential in asymmetrical 280:100 mM KCl (b). The permeability of univalent cations was estimated from the zero-current potential under bi-ionic 100 mM cation chloride : 100 mM KCl and of divalent cations 100 mM cation chloride : 100 mM $BaCl_2$, using the Goldman-Hodgkin-Katz equation. The permeability of Ba^{2+} relative to K^+ was estimated from the zero-current potential under bi-ionic 100 mM $BaCl_2$: 100 mM KCl (a), using the GHK equation appropriately modified for the presence of both monovalent and divalent cations (Lewis, 1979). Activity coefficients of electrolytes were taken from Robinson and Stokes (1959). Also shown are estimates of unitary conductance ($\pm$ SE) in symmetrical 100 mM cation chloride.

Cation	E_{rev}	P_{Cl}/P_K (%)	Conductance (pS)
K^+	0.0	100	437 ± 10 (28)
Rb^+	0.0	100	453 ± 5 (4)
Cs^+	4.5	85	389 ± 8 (3)
Na^+	6.0	78	288 ± 7 (3)
Li^+	16.5	51	149 ± 5 (4)
TEA^+	> 100	< 2	ND
Ba^{2+}	5.0 a	57	203
Sr^{2+}	1.0	54	170
Ca^{2+}	5.0	46	142
Mg^{2+}	2.0	50	113
Mn^{2+}	2.0	51	44
Cl^-	-18.5 b	7	ND

PLB (Aleksandrov *et al.*, 1976; Lunevsky *et al.*, 1977, 1983; Grishchenko *et al.*, 1984). These channels are thought to mediate depolarisation-activated Ca^{2+} currents, which show time-dependent deactivation, in algae (Berestovsky *et al.*, 1987).

It has been argued that plasma membrane Ca^{2+} channels are permeable to K^+, based on the competitive effects of external K^+ on Ca^{2+} fluxes (Felle, 1991). In addition, the 500 pS channel was inhibited by ruthenium red but not by La^{3+}, which is diagnostic of one sub-set of Ca^{2+} channels present in the plasma membrane of *Zea mays* roots, as determined by $^{45}Ca^{2+}$ uptake into vesicles isolated by two-phase partitioning (Marshall, 1991) and of Ca^{2+}-currents *in vivo* (see Tester, 1990).

The 49 pS channel was selective for monovalent cations, with the sequence Rb $\geq$ K > Cs > Na > Li, but had a low permeability for Cl⁻ (about 5% K^+) and less for divalent

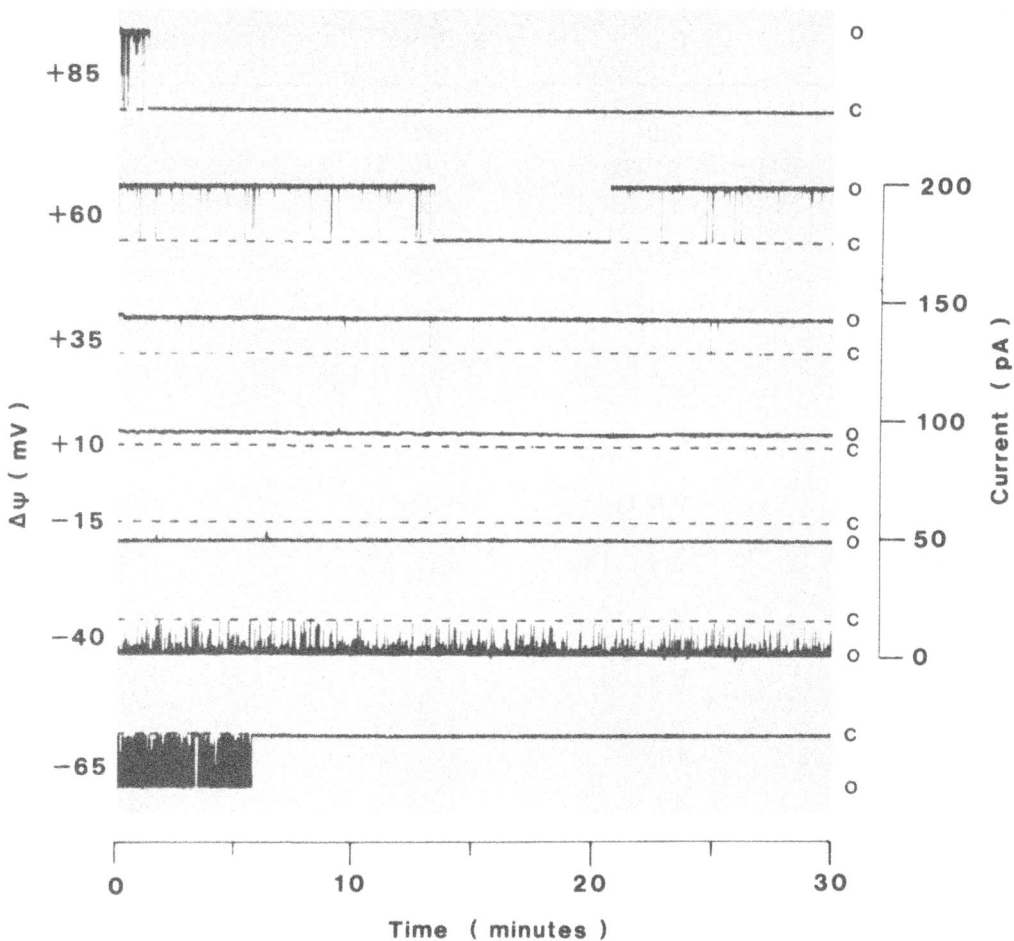

Figure 3 Single channel recordings of a 500 pS cation channel from rye root plasma membrane in symmetrical 100 mM KCl. Records were filtered mechanically at 500 Hz.

cations (White and Tester, 1991). The unitary conductance of the channel exhibited saturation kinetics with respect to KCl concentration, with a K_m KCl activity of 63 mM and a V_{max} of 79 pS (White and Tester, 1991). The gating of the channel was not voltage-dependent and the channel was open for approximately 80% of the time (Figure 4, inset). This is probably not the case *in vivo*, but currently we are unaware of the *in vivo* controls on channel gating. In symmetrical KCl concentrations the unitary conductance of the channel exhibited considerable rectification at both higher positive and higher negative voltages (Figure 4). The I/V relationship was also asymmetrical, the unitary current being greater at positive voltages (Figure 4). Thus, the 49 pS channel would mediate an inward-rectified K^+ current *in vivo*.

The 194 pS K^+ channel was permeable to K^+ but not Cl^- (White and Tester, 1991). The channel exhibited complex voltage-dependent gating (Figure 5). Channel closures at low negative membrane potentials were rare, but increased at more extreme voltages. Channel closures were more frequent at positive voltages and, as membrane potential was increased,

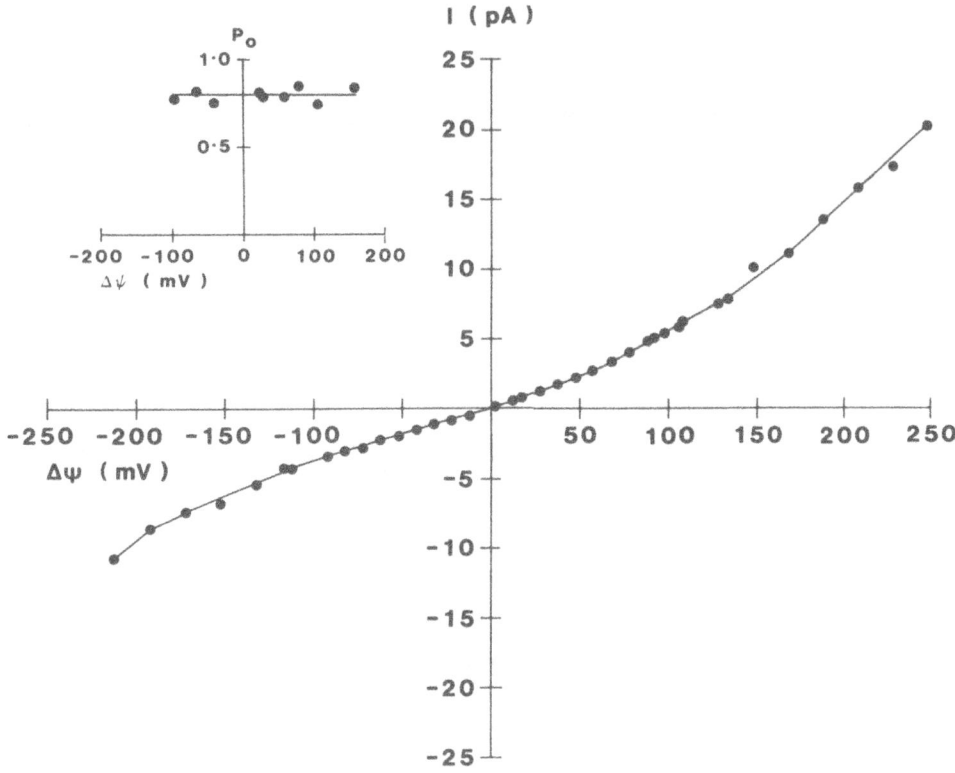

Figure 4 Unitary current / voltage relationship for the 49 pS K^+ channel from the plasma membrane of rye roots determined in symmetrical 100 mM KCl. <u>Inset</u>: Mean open probability of the 49 pS K^+ channel in symmetrical 100 mM KCl, determined from single channel records clamped at respective voltages for 30 minutes.

the frequency of transitions between open and closed states increased up to approximately +80 mV, then dropped abruptly (Figure 5). This, the channel open probability (P_o) decreased from near unity at negative membrane potentials to a minimum at about +80 mV, then increased at more extreme voltages. However, since the duration of the channel closures were extremely short at all membrane potentials, the absolute variation in P_o with membrane potential was less than 10%.

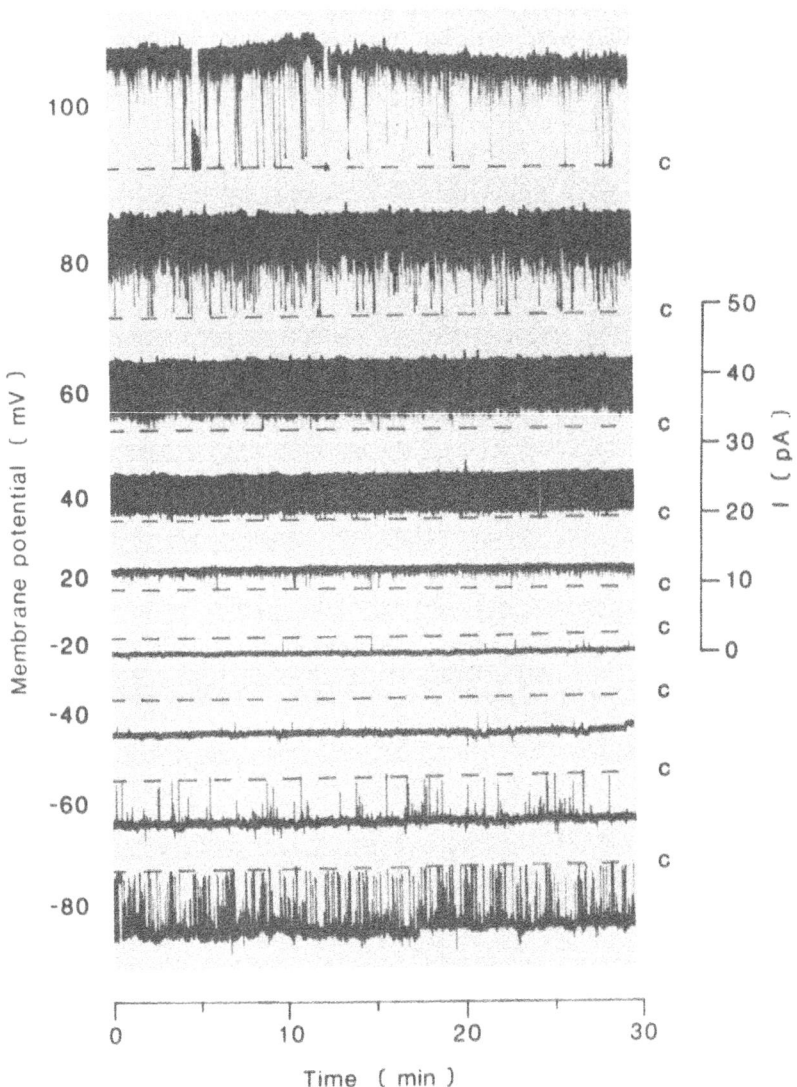

Figure 5 Single channel recordings of a 194 pS K$^+$ channel from rye root plasma membrane in symmetrical 100 mM KCl. Records were filtered mechanically at 500 Hz.

5. Future Applications of the PLB

Consistent with the advantages of the PLB technique (Section 3), we foresee the application of this technique to the study of endomembrane ion channels, in particular those of the endoplasmic reticulum, and reconstituted ion channels, especially in the elucidation of structure/function relationships (cf. Catterall, 1988; Kreuger, 1989).

One instance for which the PLB would be extremely useful would be in the study of channels associated with endomembrane systems, in particular those involved in signal transduction. At present there is much speculation about the intracellular location of the Ca^{2+} flux involved in signal transduction. Although IP_3-dependent Ca^{2+}-channels have been demonstrated at the tonoplast by patch-clamp techniques (Alexandre, Lassalles, and Kado, 1990), it is probable that Ca^{2+} channels in the endoplasmic reticulum (ER) also serve a role in regulating cytosolic Ca^{2+}, as is well established in animal systems (Carafoli, 1987; Tsien and Tsien, 1990). Thus, the need to study the channels present in the ER. In addition, the PLB provides an ideal system to study the electrical characteristics of purified endomembrane receptor-mediated channels, as illustrated by studies on the ryanodyne-sensitive (Hymel, Inui, Fleischer, and Schindler, 1988) and IP_3-sensitive (Vassilev, Kanazirska, and Tien, 1987; Ehrlich and Wattras, 1988; Bezprozvanny, Watras, and Erlich, 1991) endomembrane Ca^{2+} channels of animal cells.

The relationships between ion channel structure and function can best be studied using the combined approaches of molecular genetics, in particular the technique of site-directed mutagenesis, direct molecular characterisation and single-channel recordings (Catterall, 1988; Kreuger, 1989). This is already underway for small peptides (e.g. Brodey, Rainey, Tester, and Johnstone, 1991). Thus, the deductions of 3-dimensional protein structure and function of ion channels based on molecular biological techniques can be tested directly using biochemical, immunological and electrophysiological methods. The PLB technique would be an excellent candidate for the necessary single channel studies.

Acknowledgements

Philip White was supported by a grant from the British Science and Engineering Research Council Membrane Initiative (GR/F 33971) to Professor E.A.C. MacRobbie and Mark Tester by the Australian Research Council and a DITAC research travel grant. We thank Dr. J. Marshall (University of York, UK) for the line drawings used in this paper.

REFERENCES

ALEKSANDROV, A.A., BERESTOVSKY, G.N., VOLKOVA, S.P., VOSTRIKOV, I.Y., ZHERELOVA, O.M., KRAVCHIK, S., and LUNEVSKY, V.Z., 1976. Reconstruction of a single calcium-sodium channel of a cell in a lipid bilayer. *Doklady Akademy Nauk SSSR*, **227**, 37-40.

ALEXANDRE, J., LASSALLES, J.P., and KADO, R.T., 1990. Opening of Ca^{2+} channels in isolated red beet root vacuole membrane by inositol 1,4,5-trisphosphate. *Nature*, **343**, 567-70.

ALVAREZ, O., 1986. How to set up a bilayer system. In *Ion channel reconstitution*. Ed C. Miller. Plenum Press, New York. pp 115-130.

BERESTOVSKY, G.N., ZHERELOVA, O.M., and KATAEV, A.A., 1987. Ionic channels in characean algal cells. *Biophysics*, **32**, 1101-1120.

BEZPROZVANNY, I., WATRAS, J., and EHRLICH, B.E., 1991. Bell-shaped calcium-response curves of Ins(1,4,5)P_3- and calcium-gated channels from endoplasmic reticulum of cerebellum. *Nature*, **351**, 751-754.

BRODEY, C.L., RAINEY, P.B., TESTER, M., and JOHNSTONE, K., 1991. Bacterial blotch disease of the cultivated mushroom is caused by an ion channel forming lipodepsipeptide toxin. *Molecular Plant-Microbe Interactions*, 4, 407-411.

CARAFOLI, E., 1987. Intracellular calcium homeostasis. *Annual Review of Biochemistry*, 56, 395-433.

CATTERALL, W.A., 1988. Structure and function of voltage-sensitive ion channels. *Science*, 242, 50-61.

COHEN, F.S., 1986. Fusion of liposomes to planar bilayers. In *Ion channel reconstitution*, Ed C. Miller. Plenum Press, New York. pp 131-139.

COHEN, F.S., ZIMMERBERG, J., and FINKELSTEIN, A., 1980. Fusion of phospholipid vesicles with planar phospholipid bilayer membranes. *Journal of General Physiology*, 75, 251-270.

COHEN, F.S., NILES, W.D., and AKABAS, M.H., 1989. Fusion of phospholipid vesicles with a planar membrane depends on the membrane permeability of the solute used to create the osmotic pressure. *Ibid*, 93, 201-210.

COLOMBINI, M., 1987. Characterization of channels isolated from plant mitochondria. *Methods in Enzymology*, 148, 465-475.

CORONADO, R., 1986. Recent advances in planar phospholipid bilayer techniques for monitoring ion channels. *Annual Review of Biophysics and Biophysical Chemistry*, 15, 259-277.

EHRLICH, B.E., and WATRAS, J., 1988. Inositol 1,4,5-trisphosphate activates a channel from smooth muscle sarcoplasmic reticulum. *Nature*, 336, 583-586.

FELLE, H., 1991. Aspects of Ca^{2+} homeostasis in *Riccia fluitans*: reactions to perturbations in cytosolic-free Ca^{2+}. *Plant Science*, 74, 27-33.

FLÜGGE, U.I., and BENZ, R., 1984. Pore-forming activity in the outer membrane of the chloroplast envelope. *FEBS Letters*, 169, 85-89.

GRABOV, A.M., 1990. Voltage-dependent potassium channels in the root hair plasmalemma. *Soviet Plant Physiology*, 37, 242-247.

GRISHCHENKO, V.M., ALEKSANDROV, A.A., and BERESTOVSKY, G.N., 1984. Isolation of a fraction of cytoplasmic proteins possessing channel-forming activity from characeous algae. *Ibid*, 31, 787-793.

HANKE, W., 1986. Incorporation of ion channels by fusion. In *Ion channel reconstitution*. Ed C. Miller. Plenum Press, New York. pp 141-153.

HEDRICH, R., BARBIER-BRYGOO, H., FELLE, H., FLÜGGE, U.I., LÜTTGE, U., MAATHUIS, F.J.M., MARX, S., PRINS, H.B.A., RASCHKE, K., SCHNABL, H., SCHROEDER, J.I., STRUVE, I., TAIZ, L., and ZIEGLER, P., 1988. General mechanisms for solute transport across the tonoplast of plant vacuoles: a patch-clamp survey of ion channels and proton pumps. *Botanica Acta*, 101, 7-13.

HEDRICH, R., and NEHER, E., 1987. Cytoplasmic calcium regulates voltage-dependent ion channels in plant vacuoles. *Nature*, 329, 833-836.

HEDRICH, R., and SCHROEDER, J.I., 1989. The physiology of ion channels and electrogenic pumps in higher plants. *Annual Review of Plant Physiology and Plant Molecular Biology*, 40, 539-569.

HEDRICH, R., STOECKEL, H., and TAKEDA, K., 1990. Electrophysiology of the plasma membrane of higher plant cells: New insights from patch-clamp studies. In *The plant plasma membrane. Structure, function and molecular biology*. Eds C. Larsson and I.M.Møller. Springer-Verlag, Berlin. pp 182-202.

HILLE, B., 1984. *Ionic channels of excitable membranes*. Sinauer Associates Inc., Sunderland, Massachusetts.

HYMEL, L., INUI, M., FLEISCHER, S., SCHINDLER, H., 1988. Purified ryanodyne receptor of skeletal muscle sarcoplasmic reticulum forms Ca^{2+}-activated oligomeric Ca^{2+} channels in planar bilayers. *Proceedings of the National Academy of Sciences, USA*. 85, 441-445.

KREUGER, B.K., 1989. Toward an understanding of structure and function of ion channels. *FASEB Journal*, 3, 1906-1914.

LEWIS, C.A., 1979. Ion concentration dependence of the reversal potential and the single channel conductance of ion channels at the frog neuromuscular junction. *Journal of Physiology*, 286, 417-445.

LORD, J.M., 1983. Endoplasmic reticulum and ribosomes. In *Isolation of membranes and organelles from plant cells*. Eds J.L. Hall and A.L. Moore. Academic Press, London. pp 119-134.

LUNEVSKY, V., ALEKSANDROV, A., BERESTOVSKY, G., VOLKOVA, S., VOSTRIKOV, I., ZHERELOVA, O., 1977. Ionic mechanism of excitation of plasmalemma and tonoplast of characean algal cells. In *Transmembrane ionic exchanges in plants*. *Colloque du CNRS*, 258. Eds M. Thellier, A. Monnier, M. Demarty, and J. Dainty. pp 167-72.

LUNEVSKY, V., ZHERELOVA, O.M., VOSTRIKOV, I.Y., BERESTOVSKY, G.N., 1983. Excitation of *Characeae* cell membranes as a result of activation of calcium and chloride channels. *Journal of Membrane Biology*, **72**, 43-58.

MARSHALL, J., 1991. Calcium fluxes at the plasma membrane of *Zea mays* roots. D.Phil. Thesis, University of York.

MILLER, C., 1986. Ed *Ion channel reconstitution*. Plenum Press, New York.

MILLER, C., ARVAN, P., TELFORD, J.N., and RACKER, E., 1976. Ca^{++}-induced fusion of proteoliposomes: dependence on transmembrane osmotic gradient. *Journal of Membrane Biology*, **30**, 271-282.

MONTAL, M., and MUELLER, P., 1972. Formation of bimolecular membranes from lipid monolayers and a study of their electrical properties. *Proceedings of the National Academy of Sciences, USA.* **69**, 3561-3566.

NILES, W.D., and COHEN, F.S., 1987. Video fluorescence microscopy studies of phospholipid vesicle fusion with a planar phospholipid membrane. *Journal of General Physiology*, **90**, 703-735.

NILES, W.D., COHEN, F.S., and FINKELSTEIN, A., 1989. Hydrostatic pressures developed by osmotically swelling vesicles bound to planar membranes. *Ibid*, **93**, 211-244.

ROBINSON, R.A., and STOKES, R.H., 1959. *Electrolyte solutions*. Butterworths Scientific Publications, London.

ROUSSEAU, E., and BRISKIN, D.P., 1989. Monovalent ion selective channels derived from red beet endoplasmic reticulum. *Biophysics Journal*, **55**, 155a.

SAKMANN, B., and NEHER, E., 1983. Eds *Single channel recording*. Plenum Press, New York.

SCHINDLER, H., and ROSENBUSCH, J.P., 1978. Matrix protein from *Escherichia coli* outer membrane forms voltage-controlled channels in lipid bilayers. *Proceedings of the National Academy of Sciences, USA.* **75**, 3751-3755.

SCHONKNECHT, G., HEDRICH, R., JUNGE, W., and RASCHKE, K., 1988. A voltage-dependent chloride channel in the photosynthetic membrane of a higher plant. *Nature*, **336**, 589-592.

SMACK, D.P., and COLOMBINI, M., 1985. Voltage-dependent channels found in the membrane fraction of corn mitochondria. *Plant Physiology*, **79**, 1094-1097.

TANIFUJI, M., SATO, M., WADA, Y., ANRAKU, Y., and KASAI, M., 1988. Gating behaviours of a voltage-dependent and Ca^{2+}-activated cation channel of yeast vacuolar membrane incorporated into planar lipid bilayer. *Journal of Membrane Biology*, **106**, 47-55.

TESTER, M.A., 1988. Studies of ion channels in *Chara corallina*. Ph.D. Thesis, University of Cambridge.

TESTER, M.A., 1990. Plant ion channels: whole-cell and single-channel studies. *New Phytologist*, **114**, 305-340.

TESTER, M.A., and BLATT, M.R., 1989. Direct measurement of K^+ channels in thylakoid membranes by incorporation of vesicles into planar lipid bilayers. *Plant Physiology*, **91**, 249-252.

TESTER, M.A., and HARVEY, H.J., 1989. Verapamil-binding fraction forms Ca^{2+} channels in planar lipid bilayers. In *Plant membrane transport: The current position*. Eds J. Dainty, M.I.De Michelis, E. Marrè and F. Rasi-Caldogno. Elsevier, Amsterdam. pp 277-278.

TSIEN, R.W., and TSIEN, R.Y., 1990. Calcium channels, stores and oscillations. *Annual Review of Cell Biology*, **6**, 715-760.

VASSILEV, P.M., KANAZIRSKA, M.P., and TIEN, H.T., 1987. Ca^{2+} channels from brain microsomal membranes reconstituted in patch-clamped bilayers. *Biochimica et Biophysica Acta*, **897**, 324-330.

WADA, Y., OHSUMI, Y., TANIFUJI, M., KASAI, M., and ANRAKU, Y., 1987. Vacuolar ion channel of the yeast, *Saccharomyces cerevisiae*. *Journal of Biological Chemistry*, **262**, 17260-17263.

WHITE, P.J., 1992a. A super-maxi cation channel from the plasma membrane of rye roots characterised following incorporation into planar lipid bilayers. I. Selectivity. *Planta*, submitted.

WHITE, P.J., 1992b. A super-maxi cation channel from the plasma membrane of rye roots characterised following incorporation into planar lipid bilayers. II. Voltage-dependent kinetics. *Planta*, submitted.

WHITE, P.J., and TESTER, M.A., 1991. Potassium channels from the plasma membrane of rye roots characterized following incorporation into planar lipid bilayers. *Planta*, in press.

WOODBURY, D.J., and HALL, J.E., 1988a. Vesicle-membrane fusion: observation of simultaneous content release and membrane incorporation. *Biophysics Journal*, **54**, 345-349.

WOODBURY, D.J., and HALL, J.E., 1988b. Role of channels in the fusion of vesicles with a planar bilayer. *Ibid*, **54**, 1053-1063.

WOODBURY, D.J., and MILLER, C., 1990. Nystatin-induced liposome fusion. A versatile approach to ion channel reconstitution into planar lipid bilayers. *Ibid*, **58**, 833-839.

133

Inositol Phospholipid-Derived Signals in Plant Cells

James M. Brosnan and **Dale Sanders**

Biology Department
University of York
York, TO1 5DD, UK

Abstract

Inositol trisphosphate elicits Ca^{2+} release across the vacuolar membrane of plants. Studies on ligand binding to the solubilised inositol trisphosphate receptor have revealed that the properties of the plant receptor appear to be very similar to those of the animal cell counterpart, despite their different intracellular locations. In the light of these findings, we consider the evidence in plant cells for an inositol phospholipid signalling pathway and contrast the properties of its elements with those of animal cells.

1.   Introduction

The temptation to fit plant cell signal transduction pathways into well established animal models is almost irresistible. Although animal models may provide a useful starting point, plant researchers run the risk of carrying too many preconceptions into their investigations. Plant cells encounter a far more varied environment than animal cells and hence their signalling systems will inevitably have different requirements (Blowers and Trewavas, 1989).

Nevertheless, having acknowledged the inherent dangers in constraining concepts at the outset, a comparative approach can be useful. Similarities and differences between plant and animal cells might be used to unravel signal transduction pathways in plants, about which we have only fragmentary information at present.

2.   The Animal Model

The inositol phospholipid signalling pathway is now a standard feature in biochemical textbooks, yet it is only since the early 1980's that evidence for phosphatidylinositol 4,5-

Transport and Receptor Proteins of Plant Membranes
Edited by D.T. Cooke and D.T. Clarkson, Plenum Press, New York, 1992

135

bisphosphate (PIP$_2$) hydrolysis as a source of second messengers has generally been accepted (Berridge and Irvine, 1984). This textbook status was attained rapidly, primarily because the pathway is present in a diverse array of cell types, from slime mould (Newell, Europe-Finner, Small, and Lill, 1988) to cerebellum (Snyder and Supattapone, 1989).

The pathway in animal cells has been reviewed recently (Berridge and Irvine, 1989; Bansal and Majerus, 1990; Rana and Hokin, 1990). Briefly, an environmental stimulus (e.g. a hormone or light) activates a specific plasma membrane receptor, which in turn interacts with a G protein. The latter then stimulates a phospholipase C (PIP$_2$ phosphodiesterase) to release two messengers. One, diacylglycerol (DAG), stays within the membrane and enhances the activity of a Ca^{2+}- and phospholipid-dependent protein kinase C (PKC). The other messenger, inositol 1,4,5-trisphosphate (InsP$_3$) is water soluble and causes the release of Ca^{2+} into the cytoplasm from an intercellular store via an InsP$_3$-gated Ca^{2+} channel. The rise in cytoplasmic free Ca^{2+} then activates Ca^{2+} binding proteins, which control the physiological response.

While the basic features of the pathway are now undisputed, there remains a great deal to be discovered. The exact nature of the InsP$_3$-sensitive Ca^{2+} pool is still a matter of debate, with current suggestions favouring a discrete organelle (the calciosome) or a portion of the endoplasmic reticulum (Krause, 1991). Studies of the oscillations in free Ca^{2+} caused by InsP$_3$ and of the relationship of these oscillations to InsP$_3$-insensitive Ca^{2+} stores, are producing models for Ca^{2+}-induced cell activation (Berridge, 1991). The metabolic fate of InsP$_3$ is also of interest in the context of the production of further cell signals, notably inositol 1,3,4,5-tetrakisphosphate (InsP$_4$). InsP$_4$ is believed to cause a replenishing of the InsP$_3$-sensitive Ca^{2+} pool via extracellular Ca^{2+} influx (Irvine, 1990a), but this scheme is still disputed (Bird, Rossier, Hughes, Shears, Armstrong, and Putney, 1991a; Irvine, 1991). Finally, a whole family of PKC-type enzymes has been discovered, including some which are activated by phosphatidylcholine-derived DAG's and are hence independent of inositol phosphates and the Ca^{2+} signal (Pelech and Vance, 1989).

Do plant cells have the basic elements required for an inositol phospholipid-based signalling pathway and, if so, how do their properties compare with those of animal cells?

3.   Receptor-Linked Inositol Phospholipid Breakdown

The environmental stimulus, according to biochemical dogma, conveys its signal to the cell interior by interacting with a plasma membrane receptor. However, very little is known about plant cell receptors. Hopefully, this situation should change with the purification of auxin binding proteins and the demonstration that impermeant auxin/protein conjugates elicit auxin responses across the plasma membrane, presumably via a plasma membrane receptor (Napier and Venis, 1991).

The function and indeed the existence of receptor-coupled G proteins in plants is not clear. There are now several reports identifying G proteins from plant tissue (Drobak, Allan, Comerford, Roberts, and Dawson, 1988a; Blum, Hinsch, Schultz, and Weiler, 1988; Jacobs, Thelen, Farndale, Astle, and Rubery, 1988). These proteins have the required specificity for GTP and in some cases cross-react with antibodies raised to animal G proteins. However, the cellular events controlled by these proteins are largely a matter for speculation.

One obvious G protein target by analogy with animal cells would be phospholipase C. This enzyme activity has been convincingly demonstrated in plant cells and provides the first solid evidence for a plant phospholipid signalling pathway. Plasma membrane-associated phospholipase C shows a greater substrate preference for PIP_2 and phosphatidyli-nositol (4) phosphate (PIP) than for phosphatidylinositol (PI) (Melin, Sommarin, Sandelius, and Jergil, 1987; McMurray and Irvine, 1988). The oat root phospholipase C exhibits relative reaction rates of 10:4:1 for PIP_2, PIP and PI respectively and, therefore, the plant enzyme shows the necessary substrate specificity for a physiological role in cell signalling (Tate, Schaller, Sussman, and Crain, 1989). A common feature shared by animal and plant phospholipase C is stimulation by Ca^{2+} (see above citations). Maximum stimulation is seen at 10^{-4} M Ca^{2+} but significant increases in activity are seen over the physiological concentration range (10^{-7}-10^{-5} M), so this might present an *in vivo* control mechanism.

Reports about possible G protein activation of plant phospholipase C are conflicting. On the one hand Melin *et al.* (1987) and McMurray and Irvine (1988) could detect no effect of GTP or GTPγS on phospholipase C. However, in *Dunaliella saline* L., a rather high concentration of GTPγS (0.1 mM) did stimulate phospholipase C (Einspahr, Peeler, and Thompson, 1989). The Scottish legal idiom, *not proven*, is thus appropriate!

McMurray and Irvine (1988) found phospholipase C activity in a wide range of plant tissues and species. Interestingly, the activities varied such that celery stems have 40 times the phospholipase C activity of spinach leaves which indicates that, although widespread, phospholipase C-derived signalling systems are not likely to be of equal importance in all cell types.

For inositol phospholipids to have a role in plant cell signal transduction they must be shown to be broken down in response to the first messenger, the environmental stimulus. This criterion has been fulfilled for auxin, light and osmotic shock, if the evidence for the turnover of PIP_2 and its precursor, PIP, is taken at face value. Auxin has the ability to trigger cell division in arrested *Catharanthus roseus* L. cell cultures (Ettlinger and Lehle, 1988). When auxin is added to cells labelled with [^{3}H]inositol there is a transient fall in detectable PIP and PIP_2, accompanied by a similar rise in their hydrolysis products, $InsP_2$ and $InsP_3$. Similar results were found in the legume *Samanea saman* L. where light controls leaf movement (Morse, Crain, and Satter, 1987). Here, phosphoinositide breakdown in the leaf motor organ, monitored by [^{3}H]inositol labelling, showed a rapid decline in PIP (72%) and PIP_2 (40%) when exposed to white light. A rise in the levels of $InsP_2$, $InsP_3$ and DAG over the same timescale has also been seen (Morse, Crain, Cote, and Satter, 1989). In the unicellular green alga, *Dunaliella salina* L. osmotic shock induces degradation of PIP and PIP_2 and an increase in the amounts of phosphatidic acid, a phospholipid metabolite (Einspahr *et al.*, 1989).

These results rely on unambiguous identification of PIP_2, which is not a trivial task in plant tissue (Irvine, 1990b). Early work on PIP_2 and PIP in plants indicated that they occurred at 1-2% of the amounts of total PI seen in animal cells and thus at a sufficient content for signal transduction (reviewed in Boss, 1989). The experiments (as those outlined above) involve labelling tissue with [^{3}H]inositol or [^{32}P]phosphate and then analysing incorporation of label into lipids by TLC. However, other lipids with near identical chromatographic properties are also present in plant cells, including inositol lysophospholipids, and have led to an overestimation of PIP_2 (Drobak, Fergusion, Dawson, and Irvine, 1988b; Wheeler and Boss, 1987). Furthermore, if alkaline alcohol deacylation is employed as an analytical method, glycerophospho-inositols can produce artifacts by splitting to give inositol phosphates, which may appear to be derived from PIP_2 (Irvine, 1986). When the possible interference by these other lipids is eliminated, for example by using co-migration of PIP_2 with an authenticated internal standard, only very small amounts of PIP_2 (<0.1%) are found in higher plants (Cote, De Pass, Quarmby, Tate, Morse, Satter and Crain, 1989; Irvine, Letcher, Lander, Drobak, Dawson, and Musgrave, 1989). However, because of the sensitivity of PIP_2 to phospholipases, these observed small amounts might not be a fair reflection of the *in vivo* concentrations. Nevertheless, it would seem that only in *Chlamydomonas* (hardly a typical plant) have amounts of PIP_2 been unambiguously shown to equal those found in animal cells (Irvine *et al.*, 1989).

4.   Activation of Protein Kinases

According to the animal model, PIP_2 breakdown released two messengers: one ($InsP_3$) raises intercellular Ca^{2+}, whilst the second (DAG) stimulates a PKC. Similar DAG-responsive protein kinase activity in plants would be good, albeit circumstantial, evidence for a PIP_2-derived signalling pathway. Animal PKC's are dependent on Ca^{2+} and phospholipid and several plant protein kinases have been identified with similar properties (Harmon, 1990). However, it is not clear what the role of DAG is in the activation of these kinases. There is only one report, using *Amaranthus tricolor* L. extracts, of DAG (specifically, diolein) increasing the sensitivity of an impure plasma membrane associated kinase to Ca^{2+} (Elliot and Kokke, 1987). Phorbol esters are activators of animal PKC yet no specific binding of these agents could be found in soybean or tobacco callus, despite the presence of phospholipid-dependent protein kinases (Elliot and Skinner, 1986).

Western blotting of the *Amaranthus tricolor* L. extract showed cross-reactivity with antibodies raised to animal PKC (Elliot and Kokke, 1987). Further support for the presence of PKC-like enzymes in plants comes from screening studies on higher plant cDNA libraries of both dicots and monocots, which found sequence homologies to animal protein kinase catalytic domains (Lawton, Yamamoto, Hanks, and Lamb, 1989). However, there appears to be no conservation in putative regulatory domains away from the active site. Thus, it now seems much more likely that these plant protein kinases are regulated by lipid factors other than DAG. Of particular current interest is the potential role of lysophospholipids, which are hydrolysis products of phospholipases A_1, A_2 and B, in activating protein kinases (Martiny-Baron and Scherer, 1989). Such a kinase could account for lysophospholipid stimulation of the plasma membrane H^+-ATPase, an enzyme activity believed to be associated with the auxin-induced growth response (Morré, 1990). Given that auxin may activate a phospholipase A_2 in zucchini hypocotyls, lysophospholipids could form a signal transduction pathway in their own right (Scherer and Andre, 1989). It should also be noted that PIP and PIP_2 have also been reported to stimulate the plasma membrane H^+-ATPase (Memon and Boss, 1990).

Phospholipid- and Ca^{2+}-dependent protein kinases are likely to have a major role in plant cell signalling. However, there is little evidence to link plant PKC activity to PIP_2 breakdown.

5.   Release and Metabolic Fate of $InsP_3$

The identification of the cytosolic messenger, $InsP_3$, in plant tissue is subject to problems similar to those encountered with its precursor, PIP_2 (Irvine 1990b). Analysing the soluble fractions of [^3H]inositol-labelled plant cells using anion exchange methods established for animal cells is not practical. In plant cells, inositol is at the centre of a complex metabolic array which includes incorporation into cell wall components, many of which have comparable charges to $InsP_3$ (Cote, Quarmby, Satter, Morse, and Crain, 1990). Thus, the evidence for light-induced formation of InsP$_3$ and InsP$_2$ in *Samanea saman* L. (Morse *et al.*, 1987) is compromised by the fact that the inositol metabolite, glucuronic acid, is also present in the tissue (Cote, Morse, Crain, and Satter, 1987). Glucuronic acid co-elutes with inositol phosphates on anion exchange columns. In cell cultures of *Daucus carota* L. InsP$_3$ and InsP$_2$ are again apparently identified by anion exchange chromatography (Rincon, Chen, and Boss, 1989). However, when the column extracts are separated by paper electrophoresis, no co-migration with authentic InsP$_3$ and InsP$_2$ standards is seen. It appears that InsP$_3$ is either not present in plants in detectable amounts or is rapidly metabolised.

A new technique has been developed for assaying InsP$_3$ in tissues, which relies on extracted endogenous InsP$_3$ displacing bound [^{3}H]InsP$_3$ from a bovine adrenal preparation containing the InsP$_3$ receptor (Amersham Life Science Products, 1989). Displacement by known concentrations of InsP$_3$ produces a standard curve from which the amount of InsP$_3$ in the tissue can be assessed. When this assay was applied to maize roots and coleoptiles, endogenous InsP$_3$ was detected at between 0.1-1.0 nmol/g fresh weight (Aducci and Marrè, 1990). Assuming that 1 g of tissue is equal to 1 ml of water this would give a minimum intercellular InsP$_3$ concentration of 0.1-1.0 μM, which is fairly close to the 2 μM reported in unstimulated animal tissue (Bird, Oliver, Horstmann, Obie, and Putney, 1991b). The validity of this technique has not been fully tested in plants. The assay assumes that no other tissue components will compete for the InsP$_3$ binding site. Although the animal InsP$_3$ receptor shows high specificity for InsP$_3$, other ligands, e.g. ATP, GTP and 2,3-bisphosphoglycerate, complete for the InsP$_3$ binding site at comparatively high (but physiological) concentrations (Nunn and Taylor, 1990). However, Srivastava, Pines, and Jacoby (1989), reported that, in a similar assay, beet extracts do not interfere with InsP$_3$ standards, which suggests that competing ligands are not present in sufficient quantity to lead to an overestimation of InsP$_3$ (at least in beet).

Thus, InsP$_3$ has not been convincingly shown to exist in higher plants. Despite the lack of positive InsP$_3$ identification, the necessary features of InsP$_3$ metabolism have been found. This not only includes a pathway for InsP$_3$ sythesis via phospholipase C, but enzymes which could down-regulate an InsP$_3$ signal. Drobak, Watkins, Chattaway, Roberts, and Dawson (1991), have identified two possible avenues for InsP$_3$ metabolism in pea roots using HPLC. One would involve dephosphorylation to Ins(4,5)P$_2$. This is significant, as in animals, Ins(4,5)P$_2$ is the main InsP$_3$ phosphatase product, yet in peas the Ins(4,5)P$_2$ isomer is only found in small amounts. Further dephosphorylation products (inositol monophosphate and inositol) are also detected. Results from red beet confirm that InsP$_3$ can be broken down under conditions favouring phosphatase activity, but degradation is prevented in the presence of phosphatase inhibitors (Brosnan, 1991).

The second potential route for InsP$_3$ metabolism is further phosphorylation by an InsP$_3$ kinase into InsP$_4$. This compound is found in the pea root system, but it is only possible to designate it Ins(1,4,5,X)P$_4$ (Drobak *et al.*, 1991). If the plant InsP$_4$ has a role in Ca^{2+} mobilisation, as is ascribed to Ins(1,3,4,5)P$_4$ in animal cells, it has not yet been established.

6. InsP$_3$-Induced Ca^{2+} Efflux

Although the machinery for the production and regulation of InsP$_3$ has been identified, probably the best evidence for an InsP$_3$ signalling pathway in plants lies in the reports of InsP$_3$ induced Ca^{2+} release. It is through raising cytoplasmic Ca^{2+} that InsP$_3$ is able to propagate the cell signal in animal cells.

Early work established that InsP$_3$ could induce Ca^{2+} efflux from plant membrane fractions. However, the membrane origin of the Ca^{2+} store was not identified in zucchini hypocotyls (Drobak and Ferguson, 1985) and half maximal (K$_{0.5}$) Ca^{2+} release in microsomes from corn coleoptiles occurred at 8 μM InsP$_3$, which is about 10 times greater than the value found in most animal systems (Reddy and Poovaiah, 1987). Probably the most disturbing finding was that Ca^{2+} effluxes induced by exogenous InsP$_3$ from carrot protoplasts could be mimicked by InsP$_6$ (Rincon and Boss, 1987). The effect is unlikely to reflect any physiological process because inositol phosphates are membrane impermeant. However, the question raised by these initial findings have been largely addressed by subsequent work.

Biochemical studies on oat root vacuolar vesicles (Schumaker and Sze, 1987) and isolated vacuoles from *Acer* cells (Ranjeva, Carrasco, and Boudet, 1988) confirm that InsP$_3$ does indeed specifically release Ca^{2+} at concentrations equivalent to those found in animal cells. However, one InsP$_3$-sensitive Ca^{2+} store is clearly vacuolar. This conclusion is significant because it implies that plant cells possess a mobilisable intercellular pool of Ca^{2+} which is not present in animal cells. Despite the different cellular locations, the characteristics of InsP$_3$ induced Ca^{2+} release are very similar in plant and animal cells. The plant systems described above show K$_{0.5}$ values for InsP$_3$ action which are all sub-micromolar and no Ca^{2+} release by other inositol phosphates is seen. Therefore, InsP$_3$-induced Ca^{2+} efflux in plants has both the sensitivity and specificity for physiological action providing that InsP$_3$ is present at levels comparable to those in animals. Furthermore, TMB-8, which blocks animal InsP$_3$-gated Ca^{2+} channels, inhibit InsP$_3$-induced Ca^{2+} release in plants (Schumaker and Sze, 1987; Ranjeva *et al.*, 1988). In animal cells the glycosaminoglycan, heparin, is a potent competitive inhibitor of both InsP$_3$ induced Ca^{2+} release and InsP$_3$ binding to its receptor. InsP$_3$-induced Ca^{2+} release in plants is also sensitive to heparin (Brosnan and Sanders, 1990).

More detailed characterisation of the plant InsP$_3$ receptor will eventually come from ligand binding studies. Recently we have succeeded in measuring InsP$_3$-specific binding to a plant receptor (Brosnan, 1991). The InsP$_3$ binding assay employed for red beet microsomes is based on the solubilisation/PEG precipitation assay used for mouse cerebellum (Maeda, Niinobe, and Mikoshiba, 1990): ligand binding is measured as the displacement of [^{3}H]InsP$_3$ by competitive ligands. As anticipated, heparin binds with high affinity (K$_d$ $\simeq$ 300 nM), and the receptor has a high affinity (K$_d$ < 100 nM) for InsP$_3$ and high specificity over other inositol phosphates. In addition, InsP$_3$ binding is sensitive to sulphydryl reagents as is the animal InsP$_3$ receptor (Brosnan, 1991; Supattapone, Worley, Baraban, and Snyder, 1988). InsP$_3$ protects the plant receptor from inhibition by sulphydryl reagents, possibly suggesting the presence of a cysteine resident at, or near, the binding site.

The above results have been supported powerfully by electrophysiological measurements of an InsP$_3$-sensitive Ca^{2+} channel in red beet isolated vacuoles using the patch clamp technique (Alexandre, Lassalles, and Kado, 1990). The Ca^{2+} current has a K$_m$ of 220 nM for InsP$_3$, is insensitive to other inositol phosphates and is partially inhibited by TMB-8. Channel activity is voltage-dependent over the physiological range (+10 to +80 mV with reference to the cytoplasm). The single channel conductance is 30 pS (5 mM Ca^{2+} on the vacuolar side). Unfortunately, the frequency with which the InsP$_3$ sensitive channel is found via patch camp appears to be low. Other researchers have occasionally seen InsP$_3$ dependent ion fluxes but in whole-vacuole mode only (Dr Eva Johannes personal communication).

Both the biochemical and electrophysiological data emphasize the tonoplast as a site for InsP$_3$ action in plant cells. This does not preclude the possibility of InsP$_3$-induced Ca^{2+} release from the endoplasmic reticulum. Immature or secretory plant issue, where the endoplasmic reticulum is the dominant intracellular compartment, would provide good experimental systems to investigate this possibility. However, the tonoplast is likely to prove to be an important store of Ca^{2+} for mobilisation in response to cell signals. An additional Ca^{2+} channel has been found in the tonoplast by both biochemical and electrophysiological techniques (Sanders, Miller, Blackford, Brosnan, and Johannes, 1990; Johannes, Brosnan, and Sanders, 1991a, 1991b). This Ca^{2+} channel is voltage-gated and has a unique pharmacological profile, including inhibition by Gd^{3+}. The $InsP_3$ insensitive channel would provide a parallel Ca^{2+} release pathway, independent of $InsP_3$, for use in signal transduction.

The physiological responses controlled by $InsP_3$ elevation of free Ca^{2+} are uncertain. The most detailed investigations (on guard cells: Blatt, Thiel, and Trentham, 1990; Gilroy, Read, and Trewavas, 1990) demonstrate that artificial elevation of $InsP_3$ results in stomatal closure and inactivation of inwardly-rectifying K^+ channels, which is consistent with the events following treatment with abscisic acid (ABA). However, since artificial elevation of free Ca^{2+} clearly causes closure (Gilroy *et al.*, 1990) the findings relating to $InsP_3$ do not unequivocally implicate inositol lipid turnover as part of the *physiological* signal transduction pathway. Nevertheless, there are clear indications that hypertonicity in beet causes elevation of $InsP_3$ levels as part of a regulatory response which involves control of cell turgor in red beet (Srivastava *et al.*, 1989). These results suggest the mediation of free Ca^{2+} in control of cell turgor - a finding which is supported by independent work on brackish water charophyte algae (Okazaki & Tazawa, 1990).

To conclude, the balance of the evidence suggests that there is an inositol phospholipid-derived signalling pathway in plants which operates in an analogous fashion to animal cells by raising cytosolic $[Ca^{2+}]$. It is vital that the two major signalling components of the pathway, PIP_2 and $InsP_3$, are shown to occur at sufficiently high concentrations in plant cells. The relative importance of this signalling pathway to plant physiology is not known. Rapid turnover of $InsP_3$ in the cytoplasm, suggests that $InsP_3$ might be responsible only for the control of short term responses such as stomatal closure (Irvine, 1990b).

Although the inositol signalling pathway is perhaps not ready for inclusion in plant biochemistry textbooks, it should gain acceptance as one of the possible mechanisms for plant cell signal transduction.

Acknowledgements

We thank the SERC for an Earmarked studentship awarded to James Brosnan and Dr. M Tester (Botany Department, University of Adelaide) for some useful comments on the manuscript.

REFERENCES

ADDUCI, P., and MARRÈ, M., 1990. IP₃ levels and their modulation by fusicoccin measured by a novel [³H]IP₃ binding assay. *Biochemical and Biophysical Research Communications*, **168**, 1041-1046.

ALEXANDRE, J., LASSALLES, J.P., and KADO, R.T., 1990. Opening of Ca^{2+} channels in isolated red beet root vacuole membrane by inositol 1,4,5-trisphosphate. *Nature*, **343**, 567-570.

BANSAL, V.S., and MAJERUS, P.W., 1990. Phosphatidylinositol-derived precursors and signals. *Annual Review of Cell Biology*, **6**, 41-67.

BERRIDGE, M.J., 1991. Cytoplasmic calcium oscillations: A two pool model. *Cell Calcium*, **12**, 63-72.

BERRIDGE, M.J., and IRVINE, R.F., 1984. Inositol trisphosphate, a novel second messenger in cellular signal transduction. *Nature*, **312**, 315-321.

BERRIDGE, M.J., and IRVINE, R.F., 1989. Inositol phosphates and cell signalling. *Nature*, **341**, 197-205.

BIRD, G.St.J., ROSSIER, M.F., HUGHES, A.R., SHEARS, S.B., ARMSTONG, D.L. and PUTNEY, J.W., 1991a. Activation of Ca^{2+} entry into acinar cells by a non-phosphorylatable inositol trisphosphate. *Nature*, **352**, 162-165.

BIRD, G.St.J., OLIVER, K.G., HORSTMAN, D.A., OBIE, J., and PUTNEY, J.W., 1991b. Relationship between the calcium-mobilising action of inositol 1,4,5-triphosphate in permeable AR4-2J cells and the estimated levels of inositol 1,4,5-trisphosphate in intact AR4-2J cells. *Biochemical Journal*, **273**, 541-546.

BLATT, M.R., THIEL, G., and TRENTHAM, D.R., 1990. Reversible inactivation of K$^+$ channels of *Vicia* stomatal guard cells following the photolysis of caged inositol 1,4,5-trisphosphate. *Nature*, **346**, 766-769.

BLOWERS, D.R., and TREWAVAS, A.J., 1989. Second messengers: Their existence and relationship to protein kinases. In *Second Messengers in Plant Growth and Development*. Eds W.F. Boss and D.J. Morré. pp 1-28. Alan R. Liss, New York.

BLUM, W., HINSCH, K.D., SCHULTZ, G. and WEILER, E.W., 1988. Identification of GTP-binding proteins in the plasma membrane of higher plants. *Biochemical and Biophysical Research Communications*, **156**, 954-959.

BOSS, W.F., 1989. Phosphoinositide metabolism: its relation to signal transduction in plants. In *Second Messengers in Plant Growth and Development*. Eds W.F. Boss and D.J. Morré. pp 29-56. Alan R. Liss, New York.

BROSNAN, J.M., 1991. Biochemical characterisation of red beet calcium channels. D.Phil. Thesis, University of York.

BROSNAN, J.M. and SANDERS, D., 1990. Inositol trisphosphate-mediated Ca^{2+} release in beet microsomes is inhibited by heparin. *FEBS Letters*, **260**, 70-72.

COTE, G.G., MORSE, M.J., CRAIN, R.C., and SATTER, R.L., 1987. Isolation of the soluble metabolites of the phosphatidylinositol cycle from *Samanea saman*. *Plant Cell Reports*, **6**, 352-355.

COTE, G.G., DePASS, A.L., QUARMBY, L.M., TATE, B.F., MORSE, M.J., SATTER, R.L., and CRAIN, R.C., 1989. Separation and characterisation of inositol phospholipids from the pulvini of *Samanea saman*. *Plant Physiology*, **90**, 1422-1428.

COTE, G.G., QUARMBY, L.M., SATTER, R.L., MORSE, M.J., and CRAIN, R.C., 1990. Extraction, separation and characterisation of the metabolites of the inositol phospholipid cycle. In *Inositol Metabolism in Plants*. Eds D.J. Morré, W.F. Boss, and F.A. Loewus. pp 113-137. Alan R. Liss, New York.

DROBAK, B.K., and FERGUSON, I.B., 1985. Release of Ca^{2+} from plant hypocotyl microsomes by inositol-1,4,5-trisphosphate. *Biochemical and Biophysical Research Communications*, **130**, 1241-1246.

DROBAK, B.K., ALLAN, E.F., COMERFORD, J.F., ROBERTS, K., and DAWSON, A.R., 1988a. Presence of guanine nucleotide-binding proteins in a plant hypocotyl microsomal fraction. *Biochemical and Biophysical Research Communications*, **150**, 899-903.

DROBAK, B.K., FERGUSON, I.B., DAWSON, A.P. and IRVINE, R.F., 1988b. Inositol containing lipids in suspension cultured cells. *Plant Physiology*, **87**, 217-222.

DROBAK, B.K., WATKINS, P.A., CHATTAWAY, J.A., ROBERTS, K. and DAWSON, A.R., 1991. Metabolism of inositol (1,4,5)trisphosphate by a soluble enzyme fraction from pea (*Pisum sativum*) roots. *Plant Physiology*, **95**, 412-419.

EINSPAHR, K.J. PEELER, T.C., and THOMPSON, G.A., 1989. Phosphatidylinositol 4,5-bisphosphate phospholipase C and phosphomonoesterase in *Dunaliella saline* membranes. *Plant Physiology*, **90**, 1115-1120.

ELLIOT, D.C., and SKINNER, J.D., 1986. Calcium-dependent, phospholipid-activated protein kinase in plants. *Phytochemistry*, **25**, 39-44.

ELLIOT, D.C., and KOKKE, Y.S., 1987. Partial purification and properties of a protein kinase C type enzyme from plants. *Phytochemistry*, **26**, 2929-2935.

ETTLINGER, C., and LEHLE, L., 1988. Auxin induces rapid changes in phosphatidylinositol metabolites. *Nature*, **331**, 176-178.

GILROY, S., READ, N., and TREWAVAS, A.J., 1990. Elevation of cytoplasmic calcium by caged calcium or caged inositol trisphosphate initiates stomatal closure. *Nature*, **346**, 769-771.

HARMON, A.C., 1990. Plant lipid-activated protein kinases. In *Inositol Metabolism in Plants*. Eds D.J. Morré, W.R. Boss, and F.A.Loewus. pp 319-334. Alan R. Liss, New York.

IRVINE, R.F., 1986. The structure, metabolism and analysis in inositol lipids and inositol phosphates. In *Phosphoinositides and Receptor Mechanisms*. Ed J.W. Putney. pp 89-107. Alan R. Liss, New York.

IRVINE, R.F., LETCHER, A.J., LANDER, D.J., DROBAK, B.K., DAWSON, A.P., and MUSGRAVE, A., 1989. Phosphatidylinositol (4,5)-bisphosphate and phosphatidyl-inositol(4)-phosphate in plant tissues. *Plant Physiology*, **89**, 888-892.

IRVINE, R.F., 1990a. Quantal Ca^{2+} release and the control of Ca^{2+} entry by inositol phosphates - a possible mechanism. *FEBS Letters*, **263**, 5-9.

IRVINE, R.F., 1990b. Messenger gets the green light. *Nature*, **346**, 700-701.

IRVINE, R.F., 1991. Calcium control and InsP$_4$. *Nature*, **352**, 115.

JACOBS, M., THELEN, M.P., FARNDALE, R.W., ASTLE, M.C., and RUBERY, P.H., 1988. Specific guanine nucleotide binding by membranes from *Cucurbita pepo* seedlings. *Biochemical and Biophysical Research Communications*, **155**, 877-881.

JOHANNES, E., BROSNAN, J.M., and SANDERS, D., 1991a. Calcium channels and signal transduction in plant cells. *BioEssays*, **13**, 331-336.

JOHANNES, E., BROSNAN, J.M., and SANDERS, D., 1991b. Parallel pathways for intercellular Ca^{2+} release from the vacuole of higher plants. *The Plant Journal*, submitted.

KRAUSE, K.H., 1991. Ca^{2+} storage organelles. *FEBS Letters*, **285**, 225-229.

LAWTON, M.A., YAMAMOTO, R.T., HANKS, S.K., and LAMB, C.J., 1989. Molecular cloning of plant transcripts encoding protein kinase homologs. *Proceedings of the National Academy of Sciences, USA*. **86**, 3140-3144.

MAEDA, N., NIINOBE, M., and MIKOSHIBA, K., 1990. A cerebellar Purkinje cell marker P$_{400}$ protein is an inositol 1,4,5-triphosphate (InsP$_3$) receptor protein. Purification and characterisation of InsP$_3$ receptor complex. *The EMBO Journal*, **9**, 61-67.

MARTINY-BARON, G., and SCHERER, G.F., 1989. Phospholipid-stimulated protein kinase in plants. *Journal of Biological Chemistry*, **264**, 18052-18059.

McMURRAY, W.C., and IRVINE, R.F., 1988. Phosphatidylinositol 4,5-bisphosphate phosphodiesterase in higher plants. *Biochemical Journal*, **249**, 877-881.

MELIN, P.M., SOMMARIN, M., SANDELIUS, A.S., and JERGIL, B., 1987. Identification of Ca^{2+}-stimulated polyphosphoinositide phospholipase C in isolated plant plasma membranes. *FEBS Letters*, **223**, 87-91.

MEMON, A.R., and BOSS, W.F., 1990. Rapid light-induced change in phosphoinositide kinases and H$^+$ATPase in plasma membrane of sunflower hypocotyls. *Journal of Biological Chemistry*, **265**, 14817-14821.

MORRÉ, D.J., 1990. Activation of phospholipase a$_2$: An alternative mechanism for signal transduction. In *Inositol Metabolism in Plants*. Eds D.J. Morré, W.F. Boss, and F.A., Loewus. pp 227-257. Alan R. Liss, New York.

MORSE, M.J., CRAIN, R.C., and SATTER, R.L., 1987. Light-stimulated inositol phospholipid turnover in *Samanea saman* leaf pulvini. *Proceedings of the National Academy of Sciences, USA*, **84**, 7075-7078.

MORSE, M.J., CRAIN, R.C., COTE, G.G., and SATTER, R.L., 1989. Light-stimulated inositol phospholipid turnover in *Samanea saman* pulvini. *Plant Physiology*, **89**, 724-727.

NAPIER, R.M., and VENIS, M.A., 1991. From auxin-binding protein to plant hormone receptor? *Trends in Biochemical Sciences*, **16**, 72-75.

NEWELL, P.C., EUROPE-FINNER, G.N., SMALL, N.V., and LIU, G., 1988. Inositol phosphates, G-proteins and ras genes involved in chemotactic signal transduction of *Dictyostelium*. *Journal of Cell Science*, **89**, 123-127.

NUNN, D.L., and TAYLOR, C.W, 1990. Liver inositol 1,4,5-trisphosphate-binding sites are the Ca^{2+}-mobilising receptors. *Biochemical Journal*, **270**, 227-232.

PELECH, S.L., and VANCE, D.E., 1989. Signal transduction via phosphatidylcholine cycles. *Trends in Biochemical Sciences*, **14**, 28-30.

RANA, R.S., and HOKIN, L.E., 1990. Role of phosphoinositides in transmembrane signalling. *Physiological Reviews*, **70**, 115-164.

RANJEVA, R., CARRASCO, A., and BOUDET, A.M., 1988. Inositol trisphosphate stimulates the release of calcium from intact vacuoles isolated from *Acer* cells. *FEBS Letters*, **230**, 137-141.

REDDY, A.S., and POOVAIAH, B.W., 1987. Inositol 1,4,5-trisphosphate induced calcium release from corn coleoptile microsomes. *Journal of Biochemistry*, **101**, 569-573.

RINCON, M., and BOSS, W.F., 1987. myo-Inositol trisphosphate mobilises calcium from fusogenic carrot (*Daucus carota* L.) protoplasts. *Plant Physiology*, **83**, 395-398.

RINCON, M., CHEN, Q., and BOSS, W.F., 1989. Characterisation of inositol phosphates in carrot (*Daucus carota* L.) cells. *Plant Physiology*, **89**, 126-132.

SANDERS, D., MILLER, A.J., BLACKFORD, S., BROSNAN, J.M., and JOHANNES, E., 1990. Cytosolic free calcium homeostasis in plants. *Current Topics in Plant Biochemistry and Physiology*, **9**, 20-37.

SCHERER, G.F., and ANDRE, B., 1989. A rapid response to a plant hormone: auxin stimulates phospholipase A$_2$ *in vivo* and *in vitro*. *Biochemical and Biophysical Research Communications*, **163**, 111-117.

SCHUMAKER, K.S., and SZE, H., 1987. Inositol 1,4,5-trisphosphate releases Ca^{2+} from vacuolar membrane vesicles of oat roots. *Journal of Biological Chemistry*, **262**, 3944-3946.

SNYDER, S.H., and SUPATTAPONE, S., 1989. Isolation and functional characterisation of an inositol trisphosphate receptor from brain. *Cell Calcium*, 10, 337-342.

SRIVASTAVA, A., PINES, M., and JACOBY, B., 1989. Enhanced potassium uptake and phosphatidylinositol-phosphate turnover by hypertonic mannitol shock. *Physiologia Plantarum*, 77, 320-325.

SUPATTAPONE, S., WORLEY, P.F., BARABAN, J.M., and SNYDER, S.H., 1988. Solubilisation, purification and characterisation of an inositol trisphosphate receptor. *Journal of Biological Chemistry*, 263, 1530-1534.

TATE, B.F., SCHALLER, G.E., SUSSMAN, M.R., and CRAIN, R.C., 1989. Characterisation of a polyphosphoinositide phospholipase C from the plasma membrane of *Avena sativa*. *Plant Physiology*, 91, 1275-1279.

WHEELER, J.J., and BOSS, W.F., 1987. Polyphosphoinositides are present in plasma membranes isolated from fusogenic carrot cells. *Plant Physiology*, 85, 389-392.

Putative L-Type Calcium Channels in Plants: Biochemical Properties and Subcellular Localisation

Raoul Ranjeva, Annick Graziana, Christian Mazars and Patrice Thuleau

Centre de Physiologie Végétale
Université P. Sabatier
"Signaux et Messages Cellulaires"
Unité Associée au CNRS n°1457
118, Route de Narbonne
F-31062, Toulouse cedex, France

1.   Introduction

It is generally recognised that many essential cellular responses are regulated by the cytoplasmic free calcium ion concentration (Poovaiah and Reddy, 1987). Changes in [Ca] levels are detected by proteins and enzymes whose activities are modified in response to altered calcium levels (Hepler and Wayne, 1985: Ranjeva and Boudet, 1987); this results in the control of metabolism, gene expression and integrated functions such as exocytosis (Braam and Davis, 1990; Steer, 1988). For these reasons, calcium is considered as an important second messenger in eukaryotic organisms.

Most cells are surrounded by solutions that contain calcium in the millimolar range. However, a low cytoplasmic concentration of calcium is crucial for higher organisms whose metabolism is driven by the breakdown of ATP. If the [Ca] is above a critical threshold, calcium and phosphate would combine to form a precipitate causing the cell death. Therefore, various calcium transporting mechanisms are involved in maintaining the steep calcium gradient between the cytosol and the exterior of the cell. A message will be sensed only by the transient variation of [Ca] from the nano- to the micromolar range and depends upon the relative rates of release and uptake processes.

Exit pathways require continuous pumping of calcium from cytosol to either the extra cellular fluid or cytoplasmic organelles which may serve as intracellular calcium storage sites (e.g. vacuoles). Active extrusion systems involve either calcium ATPase or exchangers (H/Ca and Na/Ca) and are fuelled by ATP breakdown (for a review refer to Evans, Briars, and Williams, 1991).

Transport and Receptor Proteins of Plant Membranes
Edited by D.T. Cooke and D.T. Clarkson, Plenum Press, New York, 1992

145

Whereas calcium *entry* may occur by mobilisation of intracellular calcium, it is clear that the primary entry pathway for calcium in the cell is *via* the plasma membrane. In animal cells, such a process involves particular structures referred to as calcium channels that allow calcium to flow down its electrochemical gradient. (Tsien and Tsien, 1990). These proteins are essential in animal physiology where cardiovascular and neuromuscular pathologies have been shown to be related to abnormal calcium channel functions (Hosey and Lazdunski, 1988). Physiological disorders due to calcium have been also described in plants (Hepler and Wayne, 1985; Poovaiah and Reddy, 1987) but the mechanisms involved have yet to be elucidated.

In spite of the demonstration that calcium influx systems are important in plant physiology, direct evidence for plant calcium channels was lacking, until recently (Hedrich and Schroeder, 1989). At present, the occurrence of voltage-dependent channels at the plasma membrane has still to be firmly established and insight into their structure and regulation is still fragmentary (Schroeder and Thuleau, 1991).

The purpose of this paper is to summarise current understanding of the putative plant calcium channels associated with the plasma membrane; particular emphasis will be given to their biochemical characterisation and their possible role in the perception and transduction of external stimuli.

2.   Ion Channels Allow Direct Stimulus-Response Coupling

The conversion of an external stimulus to a biological response involves different steps including (i) the initial recognition of the stimulus at the membrane level, (ii) the changes in the intracellular amounts of second messengers and (iii), the triggering of second messenger dependent activities.

Regulation by important second messengers such as cyclic AMP requires a high energy cost in that cAMP has to be synthesised and degraded each time it carries a message. This is not the case for regulatory ions that enter the cytoplasm through channels down their electrochemical gradient. For example, upon membrane depolarisation the voltage-sensitive calcium channel opens and admits approximately three thousands calcium ions per millisecond into muscle cells (Hosey and Lazdunski, 1988; Glossmann and Striessnig, 1991). Therefore, the process takes place very rapidly at low energy cost and is stopped upon repolarisation of the membrane. The fine tuning of the channel activity may be achieved by effector-dependent regulation, such as protein phosphorylation, that makes the system very flexible.

As a consequence, if a component of the channel can also act as a receptor of a biologically important effector, the generation of second messenger will take place immediately.

3.   Pharmacologically Active Drugs as Calcium Channel Probes

Different types of calcium channels have been characterised in animal cells by following the mechanisms involved in their opening and closing. Those that are voltage-regulated open upon depolarisation and close on repolarisation. A particular sub-class referred to as L-type calcium channels display some interesting properties that have enabled us to unravel their molecular and functional properties. These high-voltage activated channels conduct a long-lasting current on large conductance and interact specifically with pharmacologically active compounds that modulate their opening state. Calcium channel effectors are chemically unrelated drugs that belong to three main classes: the 1,4-

dihydropyridines, the phenylalkylamines and the benzothiasepine derivatives (representative members of these chemical classes are shown in Figure 1). Most of them inhibit calcium channels (channel blockers or antagonists). However, two particular 1,4-dihydropyridine derivatives named Bay k 8644 and CGP 28392 are stimulatory (channel agonists).

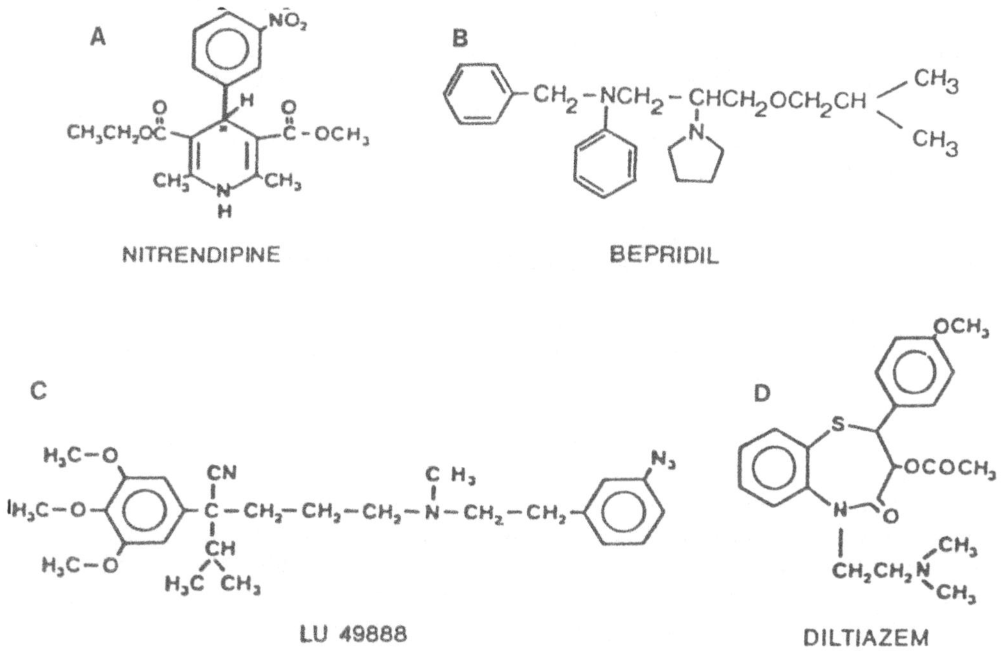

Figure 1 Antagonists for L-type calcium channels. A: 1,4 dihydropyridine; B-C: phenylalkylamines; D: benzotiazepine

These calcium channel effectors have been demonstrated to disrupt or to promote a number of plant functions. One of the most significant examples is the formation of bud initials in the moss *Funaria* (Conrad and Hepler, 1988) that is elicited by micromolar concentrations of cytokinin. However, cells cultured in cytokinin-containing medium plus a calcium channel blocker are unable to produce buds. Conversely, application of calcium channel agonist CGP 28392 induces bud initials even in the absence of the plant growth substance.

From these results, it has been suggested that the cytokinin-induced bud formation in the moss *Funaria* is due to the stimulation of calcium uptake through channels resembling the animal L-type calcium channels.

The availability of tritium-labelled derivatives for all classes of calcium channel effectors has made it possible to perform binding experiments with membranes derived from plant cells. Kinetics, affinity parameters and specificity towards the different ligands may be determined by such an approach.

The data obtained with different plant systems may be summarised as follows:

- calcium channel antagonists of the phenylalkylamine series (verapamil or desmethoxyverapamil) bind to membranes isolated from algae (Dolle and Nultsch, 1988a) and higher plants including monocots (Andrejauskas, Hertel, and Marmé, 1985; Harvey, Venis, and Trewavas, 1989) as well as dicots (Andrejauskas *et al.*,

1985; Graziana, Fosset, Ranjeva, Hetherington, and Lazdunski, 1988),
- binding sites for other classes of calcium channel blockers (e.g. diltiazem and 1,4-dihydropyridine) have been positively identified in algae only (Dolle, 1988; Dolle and Nultsch, 1988b).

A common characteristic for the phenylalkylamine binding sites is the relative high density of receptor with B_{max} ranging from 60 to 140 pmol/mg protein. Surprisingly, these figures are similar to data obtained with T-tubule membranes from skeletal muscle which are considered to be very rich in calcium channels (Hosey and Lazdunski, 1988; Glossmann and Striesnig, 1991). However, the relative affinity is in the 100 nM range for plant membranes, almost two orders of magnitude lower than for the animal counterpart.

Another intriguing feature is the apparent lack of binding sites for dihydropyridine and benzothiazepine derivatives in membranes isolated from higher plants as they are allosterically coupled and even may compete for the same sites in animals (Hosey and Lazdunski, 1988; Glossmann and Striesnig, 1991). Although it cannot be concluded that L-type like calcium channels exist in plants, these results establish that the structures of the receptor from plants are clearly different from the animal counterpart. Such an idea is strengthened by the inability of monoclonal or polyclonal antibodies recognising the calcium channel protein in animal membranes to cross-react with plant membranes (Graziana et al., 1988; Harvey et al., 1989). Moreover, differences also exist between lower and higher plants.

Concerning the specificity of the receptor, a stereoisomeric effect between enantiomeric pairs of calcium channel blockers has been observed with membranes isolated from *Daucus carota* cell cultures (Graziana et al., 1988). Thus, specific binding of (-)[^{3}H] desmethoxyverapamil was specifically inhibited by increasing concentrations of unlabelled (-) desmethoxyverapamil and other compounds of the phenylalkylamine series and bepridil. However, (-) stereoisomers were three to twenty-fold more active than (+) stereoisomers.

Binding experiments do not necessarily give information on the functional properties of the receptors. In particular, it is not clear if calcium movement is affected by the association of the calcium channel effectors to membranes. This question has been indirectly addressed by measuring calcium entry into carrot protoplasts in different conditions (Graziana et al., 1988). Briefly, protoplasts were incubated in a medium containing $^{45}Ca^{2+}$ and unlabelled calcium chloride. The label associated with the protoplasts was measured as a function of time. The uptake appears to be rapid, reaching steady-state levels within one minute. Pre-incubation of the sample in the presence of 50 μM (-) desmethoxyverapamil eliminated the calcium influx to a large extent. The process was dose-dependent and interestingly, the efficiencies of the compounds to inhibit calcium uptake coincided qualitatively, with their ability to bind membrane preparations. Thus, drugs that were ineffective in displacing the tritiated ligand from its membrane-bound receptor sites were unable to inhibit calcium uptake. In the case of active compounds, the stereoisomeric effect established for the binding experiments was also observed for calcium uptake.

Higher concentrations of drugs (3- to 10-fold) were necessary to inhibit calcium uptake rather than to complete with the binding sites on isolated membranes. This suggests that all the binding sites may not be related to calcium transport as established for animals where approximately five to ten *per cent* of the receptors are true calcium channels (Glossmann and Striessnig, 1991). By the whole-cell tight-seal technique on corn protoplasts, Ketchum and Poole (1991) have found a calcium transport system with similarities to an L-type voltage dependent calcium channel. As in the case of calcium uptake by carrot protoplasts, the process was not sensitive to diltiazem but was inhibited by phenylalkylamines and bepridil. Even if no obvious binding sites have been characterised in membrane preparation from corn (Harvey *et al.*, 1989), nifedipine (a 1,4-dihydropyridine derivative) caused inhibition of current. However, Ketchum and Poole (1991) have mentioned that the sensitivity of calcium transport to dihydropyridine varies as a function of the cell growth cycle and diminishes when the cell(s) mature(s). Conversely, MacRobbie and Banfield (1988); Tester and MacRobbie (1990) have obtained data demonstrating that only 1,4-dihydropyridines were found to inhibit and bepridil to stimulate calcium influx into *Chara coralline* and *Commelina communis* (MacRobbie, 1989). Preliminary results from our laboratory (Figure 2) show that carrot protoplasts may be loaded with the fluorescent calcium probe Fluo 3-AM.

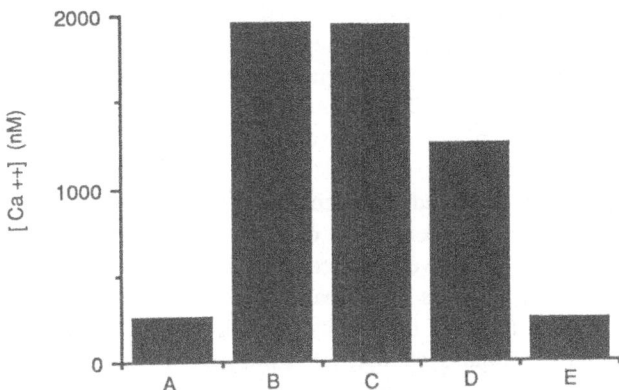

Figure 2 Cytoplasmic calcium in carrot protoplasts.
The measurements have been performed by fluorimetry using fluo3-AM as the fluorescent probe (unpublished data). A: protoplasts at resting state; B: depolarised protoplasts; C: B plus 50μM (+) bepridil: D: B plus 50 μM (-) bepridil; E: protoplasts at resting state plus 50 μM (-) bepridil.

At resting state, [Ca] was in the rang of 250 nM, as reported by other authors (for a review refer to Evans *et al.*, 1991). Upon depolarisation of plasma membrane, the cytosolic calcium rose to 1950 nM; (+) bepridil, a poor inhibitor of calcium uptake exerted a very slight inhibitory effect whereas (-) bepridil eliminated 45 *per cent* of the increase elicited by depolarisation. Such a result is consistent with earlier measurements done by Clarkson, Brownlee, and Ayling (1988) on hair root cells. Moreover, Siebers, Gräf, and Weiler (1990) have characterised calcium fluxes across purified plasma membrane of *Commelina communis*. The pharmacology of the process has not been established but the work represents the first description of a cell-free system displaying calcium channel properties in higher plants.

4. Biochemical Characterisation and Physiological Role of Putative Channel Components

In animal systems, the structure and molecular properties of L-type calcium channels have been clearly established. In particular, it has been shown that the α_1-subunit is both the receptor for calcium channel effectors and represents the calcium permeation system (Glossmann and Striessnig, 1991). In plants, only a few results have been obtained to date. Harvey *et al.*, (1989) have solubilised proteins from corn plasma membrane which retain the ability to bind verapamil. A combination of different types of chromatography resulted in a 1210-fold purification of the receptor. Analysis of the polypeptides by denaturing polyacrylamide gel electrophoresis has shown that only four bands were present. The major component was a protein of M_r 169000 resembling the animal α_1-subunit in terms of its size.

A slightly different approach has been used in our laboratory to address the question of the identification of the binding protein in carrot cell suspension cultures. In this way, and azidophenylalkylamine derivative (referred to as LU 49888) has been chosen as the ligand (Thuleau, Graziana, Canut, and Ranjeva, 1990). Binding was effective on membranes and detergent-solubilised proteins and the photoaffinity probe was covalently coupled to its potential receptor in a photo-dependent manner. The radioactive "receptor-ligand" complex was identified by fluorography as a 75-kDa peptide and was located primarily at the plasma membrane (70 *per cent*) and to a lesser extent at the tonoplast (30 *per cent*). The significance of the tonoplast associated receptor remains unclear at the present time.

Apparently, the results obtained from corn and carrot were different and functional properties of the binding protein were to be determined.

Tester and Harvey (1988) have inserted the partially purified receptor from corn coleoptile into planar lipid bilayers and measured channel currents. They have clearly established the cation selectivity of the channels. These were more permeable to barium that to calcium, as with animal systems, suggesting the existence of functional calcium channels in the preparation.

In collaboration with Dr. J. Schroeder (University of California at San Diago), proteins solubilised from plasma membranes of carrot cells have been partially purified by affinity chromatography (unpublished data). The preparation containing the 75-kDa peptide has been incorporated into phosphatidylcholine vesicles which were transformed into giant proteoliposomes. Single-channel patch clamp studies of these proteoliposomes shown calcium activities which appear to be cation selective (Thuleau, Graziana, Ranjeva, and Schroeder, 1991). One conductance state was inhibited by bepridil but more experiments are necessary before drawing definitive conclusions (Thuleau *et al.*, 1991).

As already mentioned, the direct coupling of a biologically regulatory compound to calcium channel activities is lacking in plants. However, some data have been obtained concerning the action of the aromatic phytotoxin zinniol on calcium uptake and ligand binding (Thuleau, Graziana, Rossignol, Kauss, Auriol, and Ranjeva, 1988). Radioactive zinniol was able to bind to membrane preparation from carrot cell suspension cultures in a reversible and saturable manner which suggests the existence of a receptor for the phytotoxin. Receptor occupancy stimulates the calcium uptake into protoplasts derived from the cell culture. Membranes from cell lines selected for resistance to zinniol have a very low affinity for the ligand. The phytotoxin does not significantly affect the calcium influx which remains sensitive to the inhibitory effects of bepridil. Zinniol inhibits the binding of the channel blocker in sensitive lines but have no effect in the resistant cell lines.

Such results suggest the occurrence of two distinct binding sites in the same plant. One site would be common to both calcium channel blockers and zinniol and the other one to channel blocker. This hypothesis is currently under study with reconstituted systems.

5. Conclusions and Prospects

In the field of second messenger in plants, an integrated approach involving cell biology and cell molecular biology has been achieved by Knight, Campbell, Smith, and Trewavas (1991). These authors have transformed *Nicotiana plumbaginifolia* with the gene of apoaequorin and have been able to reconstitute aequorin in whole seedlings. Since the calcium-sensitive luminescent protein is located in the cytoplasm, the transformed plants contain a constitutive calcium indicator. Therefore it has been possible to demonstrate directly (at least qualitatively) that cold-shock, touch and fungal elicitors elicit significant changes in cytosolic calcium. This clearly establishes the involvement of calcium in the transduction of various external stimuli that have been shown to modulate gene expression.

Concerning the regulation of calcium entry into the cytoplasm *via* the plasma membrane, calcium channels are likely to play a central role. This has been clearly established in animal cells. For example, an expression plasmid carrying the complementary DNA for the α_1-subunit micro-injected into cultured cells from mice with muscular dysgenesis has been shown to restore both channel activity missing from these cells and excitation-contraction coupling (Mikami, Imoto, Tanabe, Niidome, Mori, Takeshima, Narumyia, and Numa, 1989). In plants, single-channel activities have been measured on partially purified and reconstituted proteoliposomes (Tester and Harvey, 1988; Thuleau *et al.*, 1991) but their regulation remains unknown. Moreover, the exact molecular size of the channel and its subunit composition must be determined. Due to their low abundance, the isolation of channel components is difficult, even if purification procedures are available. Furthermore, in this case, since animal molecular probes do not cross-react with plants (Graziana *et al.*, 1988; Harvey *et al.*, 1989), tools specific to plants have to be constructed and used. The use of cell lines with different sensitivity to potential external signals (toxins, growth substances) that most likely interact with calcium channels is of utmost importance (Hedrich and Schroeder, 1989; Ranjeva and Boudet, 1990; Tester, 1990).

Acknowledgements

Thanks are due to Professor Michel Lazdunski and Dr. Michel Fosset who have guided us in the study of calcium channel, to Dr. Herrade Stoeckel for helpful discussions and to Dr. Deborah Goffner for checking the English version of the manuscript. This work was supported by the European Economic Community, the Centre National de la Techerche Scientifique, the French Ministère de la Recherche, and the Région Midi-Pyrénées.

REFERENCES

ANDREJAUSKAS, E., HERTEL, R., and MARMÉ, D., 1985. Specific binding of the calcium antagonist [³H]verapamil to membrane fractions from plants. *The Journal of Biological Chemistry*, **260**, 5411-5414.

BRAAM, J., and DAVIS, R.W., 1990. Rain-, wind- and touch-induced expression of calmodulin and calmodulin-related genes in *Arabidopsis*. *Cell*, **60**, 357-364.

CLARKSON, D.T., BROWNLEE, C., and AYLING, S.M., 1988. Cytoplasmic calcium measurements in intact higher plant cells: results from fluorescence ratio imaging of fura-2. *Journal of Cell Science*, **91**, 71-80.

CONRAD, P., and HEPLER, P.K., 1988. The effect of 1,4-dihydropyridines on the initiation and development gametophore buds in the moss *Funaria*. *Plant Physiology*, **86**, 684-687.

DOLLE, R., 1988. Isolation of plasma membrane and binding of the calcium antagonist nimodipine in *Chlamydomonas reinhardtii*. *Physiologia Plantarum*, **73**, 1-14.

DOLLE, R., and NULTSCH, W., 1988a. Specific binding of the calcium channel blocker [³H]verapamil to membrane fractions of *Chlamydomonas reinhardtii*. *Archives of Microbiology*, **149**, 451-458.

DOLLE, R., and NULTSCH, W., 1988b. Characterisation of D-[³H]*cis*-diltiazem binding to membrane fractions and specific binding of calcium channel blockers to isolated flagellar membranes of *Chlamydomonas reinhardtii*. *Journal of Cell Science*, **90**, 457-463.

EVANS, D.E., BRIARS, S.-A., and WILLIAMS, L.E., 1991. Active calcium transport by plant cell membranes. *Journal of Experimental Botany*, **42**, 285-303.

GLOSSMANN, H., and STRIESSNIG, J., 1991. Molecular properties of calcium channels. *Reviews in Physiology Biochemistry and Pharmacology*, **114**, 1-105.

GRAZIANA, A., FOSSET, M., RANJEVA, R., HETHERINGTON, A., and LAZDUNSKI, M., 1988. Calcium channel inhibitors that bind to plant cell membranes blocks calcium entry into protoplasts.. *Biochemistry*, **27**, 764-768.

HARVEY, H.J., VENIS, M.A., and TREWAVAS, A.J., 1989. Partial purification of a protein from maize (*Zea mays*) coleoptile membranes binding the calcium channel antagonist verapamil. *The Biochemical Journal*, **257**, 95-100.

HEDRICH, R., and SCHROEDER, J.E., 1989. The physiology of ion channels and electrogenic pumps in higher plant cells. *Annual Review of Plant Physiology and Plant Molecular Biology*, **40**, 539-569.

HEPLER, P.K., and WAYNE, R.O., 1985. Calcium and plant development. *Annual Review of Plant Physiology*, **36**, 397-439.

HOSEY, M.M., and LAZDUNSKI, M., 1988. Calcium channels: molecular pharmacology, structure and regulation. *The Journal of Membrane Biology*, **104**, 81-105.

KETCHUM, K.A., and POOLE, R.J., 1991. Cytosolic calcium regulates a potassium current in corn (*Zea mays*) protoplasts. *The Journal of Membrane Biology*, **119**, 277-288.

KNIGHT, M.R., CAMPBELL, A.K., SMITH, S.M., and TREWAVAS, A.J., 1991. Transgenic plant aequorin reports the effects of touch and cold-shock and elicitors on cytoplasmic calcium. *Nature*, **352** 524-526.

MacROBBIE, E.A.C., 1989. Calcium influx at the plasmalemma of isolated guard cells of *Commelina communis*. Effects of abscisic acid. *Planta*, **178**, 231-241.

MacROBBIE, E.A.C., and BANFIELD, J., 1988. Calcium influx at the plasmalemma of *Chara corallina*. *Planta*, **176**, 98-108.

MIKAMI, A., IMOTO, K., TANABE, T., NIIDOME, T., MORI, Y., TAKESHIMA, H., NARUMIYA, S., and NUMA, S., 1989. Primary structure and functional expression of the cardiac dihydropyridine-sensitive calcium channel. *Nature*, **340**, 230-233.

POOVAIAH, B.W., and REDDY, A.S.N., 1987. Calcium messenger in plants. *CRC Critical Reviews in Plant Sciences*, **6**, 47-103.

RANJEVA, R., and BOUDET, A.M., 1987. Phosphorylation of proteins in plants: regulatory effects and potential involvement in stimulus/response coupling. *Annual Review of Plant Physiology*, **38**, 73-93.

RANJEVA, R., and BOUDET, A.M., 1990. *Signal perception and transduction in higher plants*. Springer-Verlag, Berlin, Heidelberg, New York, London, Paris, Tokyo, Hong Kong, Barcelona.

SCHROEDER, J.I., and THULEAU, P., 1991. Calcium channels in higher plant cells. *The Plant Cell*, **3**, 555-559.

SIEBER, B., GRÄF, P., and WEILER, E.W., 1990. Calcium fluxes across the plasma membrane of *Commelina communis* L. assayed in a cell-free system. *Plant Physiology*, **93**, 940-947.

STEER, M.W., 1988. The role of calcium in exocytosis and endocytosis in plant cells. *Physiologia Plantarum*, **72**, 213-220.

TESTER, M., 1990. Plant ion channels: whole cell and single channel studies. *New Phytologist*, **114**, 305-340.

TESTER, M., and HARVEY, H.J., 1989. Verapamil-binding fractions forms calcium channels in planar lipid bilayers. In *Plant membrane transport: the current position*. Eds J. Dainty, M.I. De Michelis, E. Marrè, and R. Rasi-Rasi-Caldogno, Elsevier, Amsterdam. pp 277-278.

TESTER, M., and MacROBBIE, E.A.C., 1990. Cytoplasmic calcium affects the gating of potassium channels in the plasma membrane of *Chara corallina*: a whole-cell study using calcium-channel effectors. *Planta*, **180**, 569-581.

THULEAU, P., GRAZIANA, A., ROSSIGNOL, M., KAUSS, H., AURIOL, P., and RANJEVA, R., 1988. Binding of the phytotoxin zinniol stimulates the entry of calcium into plant protoplasts. *Proceedings of the National Academy of Sciences, USA*, **85**, 5932-5935.

THULEAU, P., GRAZIANA, A., CANUT, H., and RANJEVA, R., 1990. A 75-kDa polypeptide, located primarily at the plasma membrane of carrot cell-suspension cultures, is photoaffinity labelled by the calcium channel blocker LU49888. *Proceedings of the National Academy of Sciences, USA*, **87**, 10000-10004.

THULEAU, P., GRAZIANA, A., RANJEVA, R., and SCHROEDER, J.I., 1991. Purified calcium channel blocker binding protein from carrot cells forms calcium-permeable ions channels. *International Symposium of Plant Molecular Biology*. Tucson, in press.

TSIEN, R.W., and TSIEN, R.Y., 1990. Calcium channels, stores and oscillations. *Annual Review of Cell Biology*, **6**, 715-760.

Receptor Proteins

Introduction

The next four chapters describe some of the current research being done in the field of receptor proteins in plant membranes. It includes contributions on gibberellin and auxin receptors, as well as the 'fusicoccin receptor' and GTP-binding proteins.

The identification of plant hormone receptor proteins has been an aim which has engaged the attention of plant scientists for the past fifteen years or so. At first, progress was slow, and a number of reasons for this have been suggested. Among these are, that compared with work on the mammalian systems, plant research efforts have been very small, plant proteins are more difficult to isolate and purify and the original efforts were made by plant physiologists without the necessary back-up from biochemists.

However, over the past five years, more rapid progress has been made, due, in particular, to the adoption of a more biochemical and molecular biological approach, and using much purer membrane fractions than hitherto. These factors, along with a better understanding of the plant 'hormonal systems' and a recognition of the differences from those of animals has enabled significant advances to be made.

A major difference between plants and animals is the fact that most plant cells are totipotent. Therefore, in theory, each cell has the potential to make its own hormones, thus obviating the need for a specialised hormone secretory system. Also, an important component of the animal system, protein kinase C (PKC) has not been found in plants, although it may be possible that its function may be performed by a related PK protein with slightly different properties. From the point of view of receptor systems this may, or may not be of consequence.

Much of the recent work has focussed mainly on the plasma membrane (PM) as a site for hormone receptors, but of course other locations are known. Most notable exceptions are the endoplasmic reticulum (ER), the major site for the putative auxin receptor (auxin binding protein) and the cytosol where a polypeptide which binds gibberellin has been located. Evidence for PM surface receptors has come from work with auxin (Napier and Venis) as well as gibberellin and anti-idiotypes which bind at the cell surface with which they interact. The effect of this interaction, as well as the regulation of α-amylase gene expression at the transcription level, is fully described in the contribution from Hooley *et al*.

The best candidate for a plant hormone receptor, identified so far, has been the auxin binding protein (ABP). However, a few reservations have been expressed, in reconciling the mechanism of auxin action, particularly gene expression and cell extension, with a protein which has been located mainly in the endoplasmic reticulum (ER). Nevertheless, there may be very good reasons for this particular location, for example these ABP may be recruited by the PM during phases where sensitivity to auxin by the cell increases.

Much has been written about the fusicoccin (FC) receptor, and arguably it is the best characterised receptor in the plant kingdom. However, it still remains an intriguing question why most plants appear to have a receptor to the phytotoxic metabolite of a fungus to which only the almond (*Prunus amygdalus*) is susceptible. The mechanism by which fusicoccin (FC) stimulates the H^+-ATPase is still uncertain, and it is of great interest, in this context, how FC appears to re-activate inactive ATPases in preparations (*in vitro*).

In animal systems G-protein-coupled receptors have been identified as integral membrane proteins with seven putative membrane spanning domains, the third loop on the cytoplasmic face being involved in coupling to the G-protein. Athough, such a system has not been found in plants, GTP-binding proteins have, and their presence may lead to the identification of a G-protein-coupled receptor. G-proteins have three distinct subunits, α, β and γ, and their mode of action is as follows:

GTP binds to the α-subunit which becomes activated. This activation is followed by dissociation from the other two subunits. The activated α-subunit then modulates, for example, ion-channels or second messenger-generating enzymes. Thus, it can be seen that G-proteins represent an important component of the signal tranduction pathway.

The following contributions can be considered to represent the most recent advances in their field. However, they do pose many more questions than they answer, and as such represent an exciting challenge for the future.

Hormone Perception and Signal Transduction in Aleurone

Richard Hooley[1], Michael H. Beale[1], Sally J. Smith[1],
Robert P. Walker[1], Paul J. Rushton[1], Peter N. Whitford[1]
and Colin M. Lazarus[2]

1.   Department of Agricultural Sciences
        University of Bristol
        AFRC Institute of Arable Crops Research
        Long Ashton Research Station
        Bristol, BS18 9AF, UK

2.   Department of Botany
        University of Bristol
        Bristol BS8 1UG, UK

1.   Introduction

Aleurone cells of the Graminae are a highly specialised tissue that differentiates from peripheral endosperm cells during seed development and forms, depending on the species, either a single, or three, cell-thick layer enclosing the endosperm. When the developing seed dehydrates, endosperm cells, that contain the majority of the seed's stored carbohydrate and protein reserves, collapse and die. Aleurone cells are able to tolerate this desiccation and re-hydrate when the seed is subsequently imbibed. Shortly after the seed has germinated aleurone cells begin to synthesise and secrete a variety of hydrolytic enzymes, including α-amylase, that, together with hydrolases secreted by the scutellum epithelium, break down the stored starch and protein reserves of the endosperm to provide nutrients for the growing seedling. In addition, aleurone cells contain the majority of the seed's stored reserves of myo-inositol, phosphorous and mineral cations such as K^+ and Mg^{2+}. These are released into the endosperm after germination and provide the growing seedling with carbohydrate for cell wall synthesis, phosphorous and essential cations. Reserve mobilisation by aleurone cells appears to be coordinated to a large extent by the embryo and it is thought that one of the signals involved in this, the plant hormone gibberellin (GA), is produced by the embryo (Fincher, 1989).

Transport and Receptor Proteins of Plant Membranes
Edited by D.T. Cooke and D.T. Clarkson, Plenum Press, New York, 1992

In *Avena fatua*, GA_1 is the major biologically active GA found during germination and it appears to be synthesised by the embryo and transported into the endosperm (Hooley, Beale and Smith, 1990). The seeds germinate approximately 24-36 hours after imbibition at $25°C$. At this time, the amount of GA_1 that can be detected in the seed, after removal of the embryo, begins to increase, and by 48 hours has risen substantially. This increase in GA_1 precedes a marked rise in the amount of α-amylase detected in the embryo-less half seeds. Isolated aleurone layers of *A. fatua* respond similarly to exogenous GA_1. Amounts of α-amylase enzyme and mRNA, increase in a dose-dependent manner, reaching levels several hundred-fold higher than that in controls incubated without GA_1. GA-elevated steady-state levels of α-amylase mRNA in aleurone appear to be the result of increased transcription of α-amylase genes (Jacobsen and Beach, 1985; Zwar and Hooley, 1986). Another plant hormone, abscisic acid (ABA), accumulates in the embryo and endosperm of developing seeds of the Graminae. Although the ABA content falls as the grains dehydrate, it is present in mature seeds. ABA overcomes the effects that GA has on hydrolase gene expression in aleurone and also induces the expression of other specific genes (Fincher, 1989).

Aleurone cells are clearly able to perceive the plant hormones GA and ABA and respond to these stimuli by altering the pattern of genes that they are expressing. This article is a brief, and selective, review of the possible roles of aleurone membranes and membrane proteins in events during hormone perception, signal transduction and action in aleurone cells.

2. Uptake and Metabolism of GAs by Aleurone Cells

Some of the GA_1 released from the embryo, into the endosperm, will come into contact with the external face of the aleurone plasma membrane. What then happens to this GA_1? GAs are hydrophobic weak acids with pK_As of approximately 4.0 to 4.2 (Tidd, 1964). They can enter plant cells by passive partitioning of undissociated molecules that then become trapped as the relatively impermeable anions, in alkaline compartments within the cell (O'Neill, Keith and Rappaport, 1986). An investigation of GA_1 uptake by spinach cell suspension cultures suggests that in these cells there may be, in addition, carrier-mediated GA-transport, probably driven by a proton gradient across the plasma membrane (Nour and Rubery, 1984). At present there is no evidence to suggest that GA-transport proteins occur in aleurone plasma membrane, but this possibility has not as yet been examined in detail.

The endosperm of a germinating barley seed has a pH of approximately 5.0 (Mikola and Mikola, 1980) while the cytoplasm of a barley aleurone cell is probably about pH 7.4 (Bush and Jones, 1987). Therefore, GA_1 in the endosperm will equilibrate, and may be transported, into aleurone cells. $[^3H]GA_1$ partitions into isolated barley aleurone layers incubated at pH 4.8 (Musgrave, Kays and Kende, 1972; Nadeau, Rappaport and Stolp, 1972). However, this uptake is accompanied by metabolism of $[^3H]GA_1$ to biologically inactive GAs, particularly GA_8, by 2ß-hydroxylation, and glycosides of both GA_1 and GA_8. The available data indicate that for an aleurone cell synthesising α-amylase in response to GA_1, the concentration of GA_1 inside the cell is likely to be more than 100-fold lower than at the external face of the plasma membrane. ABA increases the rate at which $[^3H]GA_1$ is taken up by aleurone cells and at the same time increases its conversion to the inactive GA-metabolites (Nadeau *et al.*, 1972). GA uptake by plant cells is reduced substantially as the pH of the medium bathing them is increased above the pK_A (Nour and Rubery, 1984). The response of barley aleurone layers to GA_3, that has a pK_A of 3.97 (Tidd,

1964), does not change appreciably as the pH of the medium is increased over the range 4.44 to 5.25 (Goodwin and Carr, 1972). In *A. fatua* aleurone protoplasts GA_1, GA_4 and 2,2-dimethyl GA_4 are of equal biological activity (Beale, Hooley and MacMillan, 1986), even though they differ considerably in hydrophobicity (Durley and Pharis, 1972; Hoad, Phinney, Sponsel and Macmillan, 1981), and, therefore, the rate at which they are likely to cross the plasma membrane (Gimmler, Heilmann, Demmig and Hartung, 1981). Although the mechanisms and dynamics of GA transport across aleurone cell membranes are not understood in detail, these observations suggest that GA-induction of α-amylase in aleurone cells and protoplasts may not be correlated with the rate of GA-uptake.

3.   Perception of Impermeant GA_4 at the Aleurone Plasma Membrane

GA might be perceived either at the plasma membrane or within the aleurone cell, or both. We have attempted to identify the site of GA-perception in aleurone protoplasts of *A. fatua* with an impermeant derivative of GA_4 (Hooley, Beale and Smith, 1991) using a principle originally pioneered with insulin (Cautrecasas, 1969). GA_4 has the same biological activity as GA_1 in *A. fatua* aleurone protoplasts (Beale *et al.*, 1986). Analysis of the biological activities of GA_3-derivatives in cell elongation assays has indicated that elements of the D ring of the GA molecule are not of major importance in determining the biological activity of GAs (Serebryakov, Agnistikova and Suslova, 1984; Serebryakov, Epstein, Yasinskaya and Kaplun, 1984). A GA_4-derivative with a thiol-containing addition on the D-ring (GA_4-17-S-$(CH_2)_3$-SH) induces α-amylase in isolated aleurone protoplasts of *A. fatua*, although its biological activity is approximately two orders of magnitude lower than that of GA_4 (Hooley *et al.*, 1991). This derivative has been coupled, using a hydrophobic spacer arm, to 120-μm diameter beads of Sepharose 6B, such that elements of the GA_4 molecule that are most likely to confer biological activity are exposed, but can extend no further than approximately 1.95 nm from the surface of the Sepharose 6B. GA_4-17-Sepharose beads are not likely to cross the plasma membrane of aleurone cells. However, the exposed regions of the GA_4 molecule will be able to interact with both the surface of the plasma membrane, and components within the membrane to a depth of not greater than approximately 1.95nm. GA_4-17-Sepharose will not be able to penetrate the aleurone cell wall and this property has been used to monitor for any free GA_4 that might be released from the GA_4-17-Sepharose. When aleurone layers were co-incubated with aleurone protoplasts and the amounts of α-amylase mRNA induced in each tissue by the GA_4-Sepharose determined, the impermeant GA_4 was found to induce high level α-amylase gene expression in aleurone protoplasts and only very low levels of α-amylase mRNA in aleurone cells (Hooley *et al.*, 1991). These, and related observations have led us to suggest that GA_4 might be perceived at the aleurone plasma membrane and be coupled to the regulation of α-amylase gene expression by an, as yet, undefined signal transduction pathway (Hooley *et al.*, 1991).

Similar approaches have been used to localise the interaction of other ligands with plant plasma membranes. For example, aminopropyl-derivatized silica beads coated with a number of sugars have been used to try to determine the site of perception of elicitors involved in plant defense gene activation (Lienart, Gautier and Driguez, 1990). A bovine serum albumin (BSA) conjugate of periodate-oxidised deacetyl-fusicoccin binds to, and agglutinates, tobacco mesophyll protoplasts, indicating the plasma membrane location of the fusicoccin receptor (Aducci, Federico and Ballio, 1980). BSA and keyhole limpet haemocyanin, conjugates of two auxin analogues, induce auxin-like responses under conditions where they are able to interact with the plasma membrane of tobacco mesophyll protoplasts and cells of epicotyl sections from *Pisum sativum* (Venis, Thomas, Barbier-Brygoo, Ephritikhine and Guern, 1990). This adds further support to the perception of auxin at the plasma membrane (Napier and Venis, this volume).

4. Anti-Idiotypic Antibodies as Probes for Aleurone Membrane Proteins Involved in GA-Perception

GA-receptors are likely to be of low abundance in aleurone cells. We anticipate that, for a number of reasons, it will be difficult to purify and obtain sufficient receptor protein to either obtain sequence information or to raise antibodies. For example, it is not easy to isolate sufficient aleurone tissue to be able to embark with confidence on the purification of plasma membrane proteins. In addition, it is important to be able to monitor the purification of a candidate receptor using a hormone binding assay during successive stages in the purification procedure. At the present time GA-binding assays performed with aleurone microsomal membranes have not revealed any reproducible GA-binding with receptor characteristics (J.A. Napier, personal communication). When a purified receptor is not available for direct immunisation, anti-idiotypic antibodies raised against the ligand may be a source of anti-receptor antibodies, that are targeted at the ligand binding site. This anti-idiotypic approach has been used to raise antibodies specific for receptors to a variety of ligands including several mammalian hormones and neurotransmitters (Linthicum and Farid, 1988). The method has been applied to plants, and has helped to identify an integral membrane protein in the pea chloroplast envelope that may be a protein import receptor (Pain, Kanawar and Blobel, 1988), and putative receptors for auxin (Prasad and Jones, 1991) and cytokinin (Kulaeva, Karavaiko, Moshkov, Selivankina and Noivkova, 1990).

The monoclonal antibody MAC 182 (Knox, Beale, Butcher and MacMillan, 1987) shows high affinity and specific recognition of elements of the GA_4 molecule that are involved in conferring biological activity (Serebryakov *et al.*, 1984 and b). Serum from rabbits immunised with purified MAC 182 was fractionated by anion-exchange chromatography. To remove IgGs recognising the Fc region of MAC 182, IgG-containing fractions eluted from the anion-exchange column were passed through an immunoaffinity medium comprising a rat IgG of the same sub-class (IgG_{2a}) as MAC 182. Several of the partially purified IgG-containing fractions obtained compete with [^{3}H]GA_4 for binding to MAC 182. This indicates that they might contain anti-idiotypes directed against the GA_4 binding site of MAC 182. These fractions inhibit the GA-induction of α-amylase in isolated aleurone protoplasts of *A. fatua* in a dose-dependent manner, suggesting that they may be interfering with GA-perception. None of the IgGs tested are GA-agonists. In addition, they tend to cause agglutination and fusion of aleurone protoplasts, but have no appreciable effect on protoplast viability, or vacuolation. Pre-immune IgGs do not affect GA-induction of α-amylase, or protoplast agglutination. GA_1, GA_3, GA_4 and 2,2-dimethyl GA_4 are of equal biological activity in *A. fatua* aleurone protoplasts (Beale *et al.*, 1986), but have quite different affinities for MAC 182 (Knox *et al.*, 1987). The anti-idiotype shows similar antagonism of the induction of α-amylase by each of these GAs in aleurone protoplasts.

Partially purified anti-idiotypic antibodies raised against a GA_4-specific monoclonal antibody appear, therefore, to be antagonists of GA-induction of α-amylase in aleurone protoplasts and to interact with the protoplast surface. These antibodies are being used to immuno-precipitate *in vitro* translation products of aleurone mRNA.

5.   Photoaffinity Labelling of GA-Binding Proteins

A biologically active, radioiodinatable, photo-activatable analogue of GA$_4$, GA$_4$-17-yl-1'-(1'-thia)propan-3'-ol-^{125}iodo-p-azidosalicylate, ([^{125}I]GA$_4$-O-ASA) has been designed and synthesisised as a photoaffinity probe for GA-binding proteins (Beale, Hooley, Smith and Walker, 1991). It has been demonstrated that [^{125}I]GA$_4$-O-ASA is an effective GA$_4$-photoaffinity reagent and will specifically photoaffinity label MAC 182, as well as a putative GA 2ß-hydroxylase isoenzyme partially purified from cotyledons of *Phaseolus vulgaris*. We aim to characterise GA-photoaffinity labelling of aleurone plasma membrane proteins and are using dextran/PEG two-phase partitioning to try to isolate plasma membrane from aleurone protoplasts. *In vivo* photoaffinity labelling of aleurone protoplasts, combined with SDS-PAGE and fluorography, reveals that many aleurone polypeptides are photoaffinity labelled by [^{125}I]GA$_4$-O-ASA. GA$_4$ competes with the photoaffinity probe for labelling of a soluble 50 kDa polypeptide while the biologically inactive GA$_8$ does not. These labelling characteristics suggest that the 50 kDa polypeptide might be a soluble GA-receptor, an enzyme involved in GA-metabolism or a GA-receptor released from membranes during cell fractionation. Purification and characterisation of this GA-binding protein are in progress.

6.   Expression Cloning Aleurone Membrane Proteins: Strategies and Prospects

Anti-idiotypic antibodies and GA-photoaffinity probes are being used to expression clone GA-receptors that might be plasma membrane proteins. What options are available for expression cloning plant membrane proteins, and might these be applicable to cloning GA-receptors?. We have screened cDNA expression libraries constructed in lambda gt11 using an anti-idiotypic serum (Hooley, Beale, Smith and MacMillan, 1990) but have not yet isolated a candidate GA-receptor cDNA. This screening will continue with purified anti-idiotype, and will incorporate radio-iodinated GA-photoaffinity probes. Expression cloning eukaryotic membrane proteins in *Escherichia coli* is difficult, particularly if the screening procedures require, as ours do, that the protein is expressed with some functional activity, such as the capacity to bind a specific ligand. There are some successful examples of this approach. Cloned human ß$_2$-adrenergic receptor has been expressed as a ß-galactosidase fusion protein in *E. coli*. Some of the receptor molecules are cleaved from the ß-galactosidase and are inserted into the bacterial membrane, where they display ligand binding characteristics (Marullo, Delavier-Klutchko, Eshdat and Strosberg, 1988). Considerable effort has been put into the design of expression vectors, and the choice of *E. coli* strains, that favour accurate cleavage of fusion protein and targeting expressed ß-adrenergic receptors to the bacterial membrane (Marullo, Delavier-Kutchko, Guillet, Charbit, Strosberg and Emorine, 1989; Breyer, Strosberg and Guillet, 1990). It may be possible to exploit some of this technology in expression cloning plant membrane proteins in *E. coli*. However, it should be borne in mind that the likely structural similarity of mammalian G-protein coupled receptors and bacteriorhodopsin may be an important aspect of the successful expression of pharmacologically active ß-adrenergic receptors in *E. coli*.

Alternatively, it might be possible to use eukaryotic cells in expression cloning of plant membrane proteins. *Xenopus laevis* oocytes will, when injected with heterologous mRNAs, express a variety of mammalian excitatory amino acid receptors and voltage-activated ion channels, and insert them into the cell membrane where they exhibit ligand binding and electrophysiological activity (Smart, Houamed, Van Renterghem and Constantini, 1987). A cDNA cloning strategy exploiting the expression of heterologous cell membrane proteins in *X. laevis* oocytes has been devised (Masu, Nakayama, Tamaki, Harada, Kuno and Nakanishi, 1987). It involves *in vitro* synthesis of mRNA transcripts from a cDNA library, and panning through these, down to individual clones, by screening for expression of functional receptor. This technique, and an adaption of it, has been used to expression clone bovine substance-K receptor (Masu *et al.*, 1987), rat brain glutamate receptor (Hollman, O'Shea-Greenfield, Rogers and Heinmann, 1989), *Torpedo marmorata* chloride channel (Jentsch, Steinmeyer and Schwarz, 1990) and guinea-pig lung platelet-activating factor receptor (Honda, Nakamura, Miki, Minami, Watanabe, Seyama, Okado, Toh, Ito, Miyamoto and Shimizu, 1991). This approach is being explored for expression cloning the plant plasma membrane nitrate transporter (Forde, Leigh and Miller, personal communication). With suitable ligand binding, for example GA-photoaffinity labelling and/or electrophysiology assays, *X. laevis* oocytes might be suitable for cloning plant hormone receptors. Transient expression of cDNA libraries in COS cells coupled with panning, and a physical selection procedure, using antibodies to cell surface antigens, is an elegant method devised for cloning cell surface proteins (Seed, 1987; Seed and Aruffo, 1987; Simmons and Seed, 1988). In an adaption of this technique for one receptor, the need for anti-receptor antibodies has been circumvented by the use of immobilised fibroblast growth factor (Kiefer, Stephans, Crawford, Okino and Barr, 1990). Physical selection procedures clearly require high affinity interactions, and/or abundant cell surface protein. Our observation, that anti-idiotypic antibodies that may recognise GA-receptors will agglutinate aleurone protoplasts might be exploitable in a similar expression cloning strategy.

7.   Transcription Factors and Signal Transduction Pathways in Aleurone

Insights into signal perception and transduction in aleurone might be gained by identifying components of the transcriptional machinery involved in the expression of α-amylase genes. The analysis of wheat and barley α-amylase promoters fused to a reporter gene and expressed transiently in aleurone cells has defined regions of the promoters that appear to be involved in GA- and ABA-regulation of expression (Huttly and Baulcombe, 1989; Jacobsen and Close, 1991). DNase 1 footprinting has revealed that nuclear protein, extracted from GA_1-treated *A. fatua* aleurone protoplasts, binds to similar elements in the promoter regions of two wheat α-amylase genes that are regulated by GA (Rushton, Hooley and Lazarus, unpublished results). The DNase 1 footprints contain sequences similar to phorbol ester and cAMP response elements (Deutsch, Hoeffeler, Jameson and Habener, 1988). This raises the possibility that in aleurone cells transcription factors of the leucine zipper class, that may be similar to c-Fos and c-Jun of the AP-1 family (Angel, Imagawa, Chiu, Stein, Imbra, Rahmsdorf, Jonat, Herrlich and Karin, 1987; Curran and Franza, 1988) and to CREB (Montminy and Bilezikjian, 1987), might bind to these elements. The transcription factors c-Fos, c-Jun and CREB are components of signal transduction pathways activated by extracellular signals that are perceived by cell surface receptors (for a recent review see Karin, 1991). It is possible that transcription factors in aleurone that are similar to c-Fos, c-Jun and CREB may also be components of signal transduction pathways coupled to cell-surface receptors. However, there is no direct evidence to support this theory, or to suggest that GA or ABA might be involved. Nevertheless, cDNA cloning and characterisation of transcription factors involved in α-amylase gene expression in aleurone may help to identify signal transduction pathways involved in regulation of the expression of these genes. Other DNase 1 footprints in the α-amylase promoters contain sequences that are hyphenated palindromes suggesting that other types of transcription factors may also interact with the wheat α-amylase promoters.

8. Membrane Phospholipids and Gibberellin Action

Phospholipid and choline metabolism has been studied in some detail in wheat aleurone cells. During the lag phase between GA_3 addition and the induction of α-amylase, there is a marked change in the rate of turnover of N-methyl and methylene carbons of the phosphatidyl choline (PC) headgroups (Vakharia, Brearley, Wilkinson, Galliard and Laidman, 1987; Hetherington and Laidman, 1991). In non-GA_3 treated controls, N-methyl carbons turnover more rapidly than methylene carbons. GA_3 promotes phospholipid breakdown and also changes the pattern of PC metabolism such that the whole choline headgroup, rather than N-methyl carbons, turnover. It has been suggested that this change in PC turnover may be a component of a GA signal transduction pathway in aleurone (Vakharia *et al.*, 1987; Hetherington and Laidman, 1991). Changes in the phospholipid composition of aleurones from dwarf wheat varieties in response to low temperature treatment that alters the sensitivity of these cells to GA_3, have been interpreted as indicating that membrane lipids are primary sites of GA_3 perception (Singh and Paleg, 1984; 1985; 1986). In this context it may be of interest that differential scanning calorimetry and electron spin resonance spectroscopy of phospholipid liposomes indicate that a mixture of the biologically active GA_4 and GA_7 can associate with the surface of phospholipid membranes and cause disordering of the bilayer (Pauls, Chambers, Dumbroff and Thompson, 1982).

9. Calcium Status of Aleurone Cells

The calcium status of barley aleurone cells and protoplasts has been investigated by acid-loading of indo-1 and fluorescence ratio analysis, combined with studies on [^{45}Ca] transport into microsomal fractions enriched in endoplasmic reticulum (ER) (Bush and Jones, 1987, 1988; Bush, Biswas and Jones, 1989; Gilroy, Bethke and Jones, 1991). The $[Ca^{2+}]_{cyt}$ of GA_3-treated barley aleurone cells secreting α-amylase is maintained at approximately 350nM, while the $[Ca^{2+}]_{ER}$ is at least 4μM. Transport of Ca^{2+} into the ER of barley aleurone cells is several-fold higher in GA_3-treated, compared with non-GA_3 treated controls, and this appears to be due to a stimulation of the ER Ca^{2+} transporter activity by GA_3. It has been suggested that one role for the elevated $[Ca^{2+}]_{ER}$ might be to stabilise newly synthesised α-amylase, which is a Ca^{2+}-containing metalloenzyme, and this might be a point of post-translational control, by GA_3, of α-amylase. Recently, it has been reported that GA_3 increases the $[Ca^{2+}]_{cyt}$ of barley aleurone cells from approximately 100nM to 400nM prior to the stimulation in α-amylase synthesis, and that ABA overcomes this increase within 2-3 hours (Gilroy *et al.*, 1991). Ca^{2+}, and Ca^{2+}-transporting proteins, clearly play important roles in the response of aleurone cells to GA and ABA.

10. Myo-Inositol, Phosphorous and Mineral Ion Reserves in Aleurone Cells

Phytic acid (*myo*-inositol 1,2,3,4,5,6-hexa*kis*phosphate) is deposited in developing seeds as phytate, an insoluble mixed salt of several mineral cations including K, Mg, Ca and Fe (Cosgrove, 1980). Phytate is located as an inclusion in the protein bodies of oat aleurone cells (Peterson, Saigo and Holy, 1985). During germination, phytate is degraded by phytate-specific phosphohydrolases (phytases) (Loewus, Everard and Young, 1990) releasing *myo*-inositol, inorganic phosphate and the associated cations (Eastwood and Laidman, 1971). Aleurone cells sustain a considerable flux of cations and P out into the endosperm during germination. Very little is known about the mechanism underlying the flux and transport of ions across both the aleurone protein body and plasma membranes. GA_3 stimulates the efflux of K^+ ions from barley aleurone cells (Jones, 1973). The mechanism of K^+ flux across the plasma membrane of barley aleurone protoplasts has been studied using patch-clamp techniques (Bush, Hedrich, Schroeder and Jones, 1988). K^+ influx in barley aleurone protoplasts is mediated by channels, the activity of which is not greatly effected by GA_3. There is no clear evidence for the involvement of K^+ channels in the efflux of K^+ from barley aleurone protoplasts, and it seems likely that stimulation of K^+ efflux by GA_3 involves an elevation of the membrane potential. The mechanism underlying the very substantial flux of cations through the aleurone protein body and plasma membranes is open to further investigation.

11. Summary

Our understanding of the molecular mechanism of action of GA and ABA in aleurone is far from complete, and the extent to which aleurone membranes and membrane proteins are involved is just emerging. Although not abundant, aleurone is an excellent tissue for investigating hormone perception, signal transduction, gene expression and the regulation of ion transport. Prospects for progress in each of these areas are encouraging.

Acknowledgements

This research is supported by the AFRC Plant Molecular Biology Programme (grant PG 111/507 to RH and MHB) and an AFRC Link (grant LRG 115 to CLM and RH).

REFERENCES

ANGEL, P., IMAGAWA, M., CHIU, R., STEIN, B., IMBRA, R.J., RAHMSDORF, H.J., JONAT, C., HERRLICH, P., and KARIN, M., 1987. Phorbol ester-inducible genes contain a common *Cis* element recognized by a TPA-modulated *Trans*-acting factor. *Cell*, **49**, 729-739.

ADUCCI, P., FEDERICO, R., and BALLIO, A., 1980. Interaction of a high molecular weight derivative of fusicoccin with plant membranes. *Phytopathologia Mediterranea*, **19**, 187-188.

BEALE, M.H., HOOLEY, R., and MACMILLAN, J., 1986. Gibberellins: structure-activity relationships and the design of molecular probes. In *Plant growth substances 1985*. Ed M. Bopp, Springer-Verlag. pp 65-73.

BEALE, M.H., HOOLEY, R., SMITH, S.J., and WALKER, R.P., 1991. Photoaffinity probes for gibberellin-binding proteins. *Phytochemistry* (In Press).

BREYER, R.M., STROSBERG, D.A., and GUILLET, J-G., 1990. Mutational analysis of ligand binding activity of β_2 adrenergic receptor expressed in *Escherichia coli*. *EMBO Journal.*, **9**, 2679-2684.

BUSH, D.S., and JONES, R.L., 1987. Measurement of cytoplasmic calcium in aleurone protoplasts using indo-1 and fura-2. *Cell Calcium*, **8**, 455-472.

BUSH, D.S., and JONES, R.L., 1988. Cytoplasmic calcium and α-amylase secretion from barley aleurone protoplasts. *European Journal of Cell Biology*, **46**, 466-469.

BUSH, D.S., HEDRICH, R., SCHROEDER, J.I., and JONES, R.L., 1988. Channel-mediated K^+ flux in barley aleurone protoplasts. *Planta*, **176**, 368-377.

BUSH, D.S., BISWAS, A.K., and JONES, R.L., 1989. Gibberellic-acid-stimulated Ca^{2+} accumulation in endoplasmic reticulum of barley aleurone: Ca^{2+} transport and steady-state levels. *Planta*, **178**, 411-420.

CAUTRECASAS, P., 1969. Interaction of insulin with the cell membrane: the primary action of insulin. *Proceedings of the National Academy of Sciences, USA*, **63**, 450-457.

COSGROVE, D.J., 1980. *Inositol phosphates: their chemistry, biochemistry and physiology.* Elsevier Scientific Publishing Company, New York.

CURRAN, T., and FRANZA, B.R., 1988. Fos and Jun: the AP-1 connection. *Cell,* **55**, 395-397.

DEUTSCH, P.J., HOEFFLER, J.P., JAMESON, J.L., and HABENER, J.F., 1988. Cyclic AMP and phorbol ester-stimulated transcription mediated by similar DNA elements that bind distinct proteins. *Proceedings of the National Academy of Sciences, USA,* **85**, 7922-7926.

DURLEY, R.C., and PHARIS, R.P., 1972. Partition coefficients of 27 gibberellins. *Phytochemistry,* **11**, 317-326.

EASTWOOD, D., and LAIDMAN, D.L., 1971. The mobilization of macronutrient elements in germinating wheat grain. *Phytochemistry,* **10**, 1275-1284.

FINCHER, G.B., 1989. Molecular and cellular biology associated with endosperm mobilization in germinating cereal grains. *Annual Reviews of Plant Physiology and Plant Molecular Bioliogy,* **40**, 305-346.

GILROY, S., BETHKE, P., and JONES, R.L., 1991. Hormonal regulation of calcium and calmodulin in barley aleurone cells. *Plant Physiology,* **96**, Supp. Abstract 157.

GIMMLER, H., HEILMANN, B., DEMMIG, B., and HARTUNG, W., 1981. The permeability coefficients of the plasmalemma and the chloroplast envelope of spinach mesophyll cells for phytohormones. *Zeitschrift für Naturforschung,* **36**, 672-678.

GOODWIN, P.B., and CARR, D.J., 1972. The induction of amylase synthesis in barley aleurone layers by gibberellic acid I. Response to temperature. *Journal of Experimental Botany,* **23**, 1-7.

HETHERINGTON, P.R., and LAIDMAN, D.L., 1991. Influence of gibberellic acid and the *Rht3* gene on choline and phospholipid metabolism in wheat aleurone tissue. *Journal of Experimental Botany,* **42**, 1357-1362.

HOAD, G.V., PHINNEY, B.O., SPONSEL, V.M., and MACMILLAN, J., 1981. The biological activity of sixteen gibberellin A_4 and gibberellin A_9 derivatives using seven bioassays. *Phytochemistry,* **20**, 703-713.

HOLLMANN. M., O'SHEA-GREENFIELD, A., ROGERS, S.W., and HEINMANN, S., 1989. Cloning by functional expression of a member of the glutamate receptor family. *Nature,* **342**, 643-648.

HONDA, Z., NAKAMURA, M., MIKI, I., MINAMI, M., WATANABE, T., SEYAMA, Y., OKADO, H., TOH, H., ITO, K., MIYAMOTO, T., and SHIMIZU, T., 1991. Cloning by functional expression of platelet-activating factor receptor from guinea-pig lung. *Nature,* **349**, 342-345.

HOOLEY, R., BEALE, M.H., SMITH, S.J., and MacMILLAN, J., 1990. Novel affinity probes for gibberellin receptors in aleurone protoplasts of *Avena fatua.* In *Plant growth substances 1988.* Eds R.P. Pharis and S.B. Rood. Springer-Verlag. pp 145-153.

HOOLEY, R., BEALE, M.H., and SMITH, S.J., 1990. Gibberellin perception in the *Avena fatua* aleurone. In *Hormone perception and signal transduction in animals and plants.* Eds J. Roberts, C. Kirk, and M. Venis, The Company of Biologists, Cambridge. pp 79-86.

HOOLEY, R., BEALE, M.H., and SMITH, S.J., 1991. Gibberellin perception at the plasma membrane of *Avena fatua* aleurone protoplasts. *Planta,* **183**, 274-280.

HUTTLY, A.K., and BAULCOMBE, D.C., 1989. A wheat α-*Amy2* promoter is regulated by gibberellin in transformed oat aleurone protoplasts. *EMBO Journal,* **8**, 1907-1913.

JACOBSEN, J.V., and BEACH, R.L., 1985. Control of transcription of α-amylase and rRNA genes in barley aleurone protoplasts by gibberellic acid and abscisic acid. *Nature,* **316**, 275-277.

JACOBSEN, J.V., and CLOSE, T.J., 1991. Control of transient expression of chimaeric genes by gibberellic acid and abscisic acid in protoplasts prepared from mature barley aleurone layers. *Plant Molecular Biology,* **16**, 713-724.

JENTSCH, T.J., STEINMEYER, K., and SCHWARZ, G., 1990. Primary structure of *Torpedo marmorata* chloride channel isolated by expression cloning in *Xenopus* oocytes. *Nature,* **348**, 510-514.

JONES, R.L., 1973. Gibberellic acid and ion release from barley aleurone tissue. *Plant Physiology,* **52**, 303-308.

KARIN, M., 1991. Signal transduction and gene control. *Current Opinion in Cell Biology,* **3**, 467-473.

KIEFER, M.C., STEPHANS, J.C., CRAWFORD, K., OKINO, K., and BARR, P.J., 1990. Ligand-affinity cloning and structure of a cell surface heparan sulfate proteoglycan that binds basic fibroblast growth factor. *Proceedings of the National Academy of Sciences, USA,* **87**, 6985-6989.

KNOX, J.P., BEALE, M.H., BUTCHER, G.W., and MacMILLAN, J., 1987. Preparation and characterization of monoclonal antibodies which recognise different gibberellin epitopes. *Planta,* **170**, 86-91.

165

KULAEVA, O.N., KARAVAIKO, N.N., MOSHKOV, I.E., SELIVANKINA, S.Y., and NOVIKOVA, G.V., 1990. Isolation of a protein with cytokinin-receptor properties by means of anti-idiotype antibodies. *FEBS Letters*, **261**, 410-412.

LIENART, Y., GAUTIER, C., and DRIGUEZ, H., 1990. Immobilized sugars as abiotic inducers of ß-D-glycanohydrolases in plant cells. *Plant Science*, **68**, 197-202.

LINTHICUM, D.S., and FARID, N.R., 1988. *Anti-idiotypes, Receptors, and Molecular Mimicry*, Springer-Verlag.

LOEWUS, F.A., EVERARD, J.D., and YOUNG, K.A., 1990. Inositol metabolism: precursor role and breakdown. In *Inositol metabolism in plants*. Eds J.D. Morre, W.F. Boss, and F.A. Loewus. Wiley-Liss, New York. pp 21-45.

MARULLO, S., DELAVIER-KLUTCHKO, C., ESHDAT, Y., STROSBERG, A.D., and EMORINE, L., 1988. Human ß$_2$-adrenergic receptors expressed in *Escherichia coli* membranes retain their pharmacological properties. *Proceedings of the Natioal Academy of Sciences, USA*, **85**, 7551-7555.

MARULLO, S., DELAVIER-KLUTCHKO, C., GUILLET, J-G., CHARBIT, A., STROSBERG, A.D., and EMORINE, L.J., 1989. Expression of human ß1 and ß2 adrenergic receptors in *E. coli* as a new tool for ligand screening. *Biotechnology*, **7**, 923-927.

MASU, Y., NAKAYAMA, K., TAMAKI, H., HARADA, Y., KUNO, M., and NAKANISHI, S., 1987. cDNA cloning of bovine substance-K receptor through oocyte expression system. *Nature*, **329**, 836-838.

MIKOLA, L., and MIKOLA, J., 1980. Mobilization of proline in the starchy endosperm of germinating barley grain. *Planta*, **149**, 149-154.

MONTMINY, M.R., and BILEZIKJIAN, L.M., 1987. Binding of a nuclear protein to the cyclic-AMP response element of the somatostatin gene. *Nature*, **328**, 175-178.

MUSGRAVE, A., KAYS, S.E., and KENDE, H., 1972. Uptake and metabolism of radioactive gibberellins by barley aleurone layers. *Planta*, **102**, 1-10.

NADEAU, R., RAPPAPORT, L., and STOLP, C.F., 1972. Uptake and metabolism of ^{3}H-gibberellin A$_1$ by barley aleurone layers: response to abscisic acid. *Planta*, **107**, 315-324.

NAPIER, R.M., and VENIS, M.A., 1991. *This volume*.

NOUR, J.M., and RUBERY, P.H., 1984. The uptake of gibberellin A$_1$ by suspension-cultured *Spinacia oleracea* cells has a carrier-mediated component. *Planta*, **160**, 436-443.

O'NEILL, S., KEITH, B., and RAPPAPORT, L., 1986. Transport of gibberellin A$_1$ in cowpea membrane vesicles. *Plant Physiology*, **80**, 812-817.

PAIN, D., KANAWAR, Y.S., and BLOBEL, G., 1988. Identification of a receptor for protein import into chloroplasts and its localization to envelope contact zones. *Nature*, **331**, 232-237.

PAULS, K.P., CHAMBERS, J.A., DUMBROFF, E.B., and THOMPSON, J.E., 1982. Perturbation of phospholipid membranes by gibberellins. *New Phytologist*, **91**, 1-17.

PETERSON, D.M., SAIGO, R.H., and HOLY, J., 1985. Development of oat aleurone cells and their protein bodies. *Cereal Chemistry*, **62**, 366-371.

PRASAD, P.V., and JONES, A.M., 1991. Putative receptor for the plant growth hormone auxin identified and characterized by anti-idiotypic antibodies. *Proceedings of the National Academy of Sciences, USA*, **88**, 5479-5483.

SEED, B., 1987. An LFA-3 cDNA encodes a phospholipid-linked membrane protein homologous to its receptor CD2. *Nature*, **329**, 840-842.

SEED, B., and ARUFFO, A., 1987. Molecular cloning of the CD2 antigen, the T-cell erythrocyte receptor, by a rapid immunoselection procedure. *Proceedings of the National Academy of Sciences, USA*, **84**, 3365-3369.

SEREBRYAKOV, E.P., AGNISTIKOVA, V.N., and SUSLOVA, L.M., 1984a. Growth-promoting activity of some selectively modified gibberellins. *Phytochemistry*, **23**, 1847-1854.

SEREBRYAKOV, E.P., EPSTEIN, N.A., YASINSKAYA, N.P., and KAPLUN, A.B., 1984b. A mathematical additive model of the structure-activity relationships of gibberellins. *Phytochemistry*, **23**, 1855-1863.

SIMMONS, D., and SEED, B., 1988. The Fcγ receptor of natural killer cells is a phospholipid-linked membrane protein. *Nature*, **333**, 568-570.

SINGH, S.P., and PALEG, L.G., 1984. Low temperature-induced GA$_3$ sensitivity of wheat II. Changes in lipids associated with the low temperature-induced GA$_3$ sensitivity of isolated aleurone of kite. *Plant Physiology*, **76**, 143-147.

SINGH, S.P., and PALEG, L.G., 1985. Low temperature-induced GA$_3$ sensitivity of wheat. IV. Comparison of low temperature effects on the phospholipids of aleurone tissue of dwarf and tall wheat. *Australian Journal of Plant Physiology*, **12**, 277-289.

SINGH, S.P., and PALEG, L.G., 1986. Low temperature-induced GA$_3$ sensitivity of wheat. VI. Effect of inhibitors of lipid biosynthesis on α-amylase production by dwarf (*Rht3*) and tall *rht*) wheat, and on lipid metabolism of tall wheat aleurone tissue. *Australian Journal of Plant Physiology*, **13**, 409-416.

SMART, T.G., HOUAMED, K.M., VAN RENTERGHEM, C., and CONSTANTINI, A., 1987. mRNA-directed synthesis and insertion of functional amino acid receptors in *Xenopus laevis* oocytes. *Biochemical Society Transactions*, **15**, 117-122.

TIDD, B.K., 1964. Dissociation constants of the gibberellins. *Journal of the Chemical Society*, 1521-1523.

VAKHARIA, D.N., BREARLEY, C.A., WILKINSON, M.C., GAILLARD, T., and LAIDMAN, D.L., 1987. Gibberellin modulation of phosphatidyl-choline turnover in wheat aleurone tissue. *Planta*, **172**, 502-507.

VENIS, M.A., THOMAS, E.W., BARBIER-BRYGOO, H., EPHRITIKHINE, G., and GUERN, J., 1990. Impermeant auxin analogues have auxin activity. *Planta*, **182**, 232-235.

ZWAR, J.A., and HOOLEY, R., 1986. Hormonal regulation of α-amylase gene transcription in wild oat (*Avena fatua* L.) aleurone protoplasts. *Plant Physiology*, **80**, 459-463.

The Auxin Receptor: Structure and Distribution

R.M. Napier and M.A. Venis

Horticulture Research International
East Malling
West Malling
Kent, ME19 6BJ, UK

Abstract

A range of antibodies has been developed to probe the structure and function of the maize auxin receptor. The epitopes recognised by a number of polyclonal sera and our monoclonal antibodies have been mapped with the aid of an epitope mapping kit. The polyclonal sera recognise three principal epitopes, two of which are conserved in other species. Two monoclonals recognise the endoplasmic reticulum retention sequence (-KDEL) at the C-terminus, another two a region close to the N-terminus. We therefore have markers for specific parts of the receptor, and we are using these to identify functionally active parts of the protein. Additionally, we have a number of sera against a synthetic peptide. This peptide was identified from the published receptor sequence as being likely to contribute to the auxin binding site. The anti-peptide sera recognise the receptor on immunoblots and have potent auxin agonist activity on tobacco leaf mesophyll protoplasts, eliciting characteristic hyperpolarisation of the plasma membrane. By contrast, antisera to native receptor act as auxin antagonists. This auxin agonist activity means that we have identified at least part of the auxin binding site. The same experiments, and others with impermeant auxin-protein conjugates, show that the receptor is active on the outside of the plasma membrane. However, the bulk of the receptor protein is in the endoplasmic reticulum in accordance with its KDEL targeting sequence and this dual distribution will be discussed.

Keywords

Auxin receptor, antibody, epitope, endoplasmic reticulum, plasma membrane.

Transport and Receptor Proteins of Plant Membranes
Edited by D.T. Cooke and D.T. Clarkson, Plenum Press, New York, 1992

169

1. Introduction

Considerable effort has been devoted to the identification of plant hormone receptors, and this has been most successful in the case of auxin. In 1985 Löbler and Klämbt (1985a) reported the first complete purification of a putative auxin receptor (referred to in the literature as auxin-binding protein, ABP) and work on this protein has progressed rapidly since. Other purification procedures have been developed (Shimomura, Sotobayashi, Futai, and Fukui, 1986 ; Napier, Venis, Bolton, Richardson, and Butcher, 1988) and antibodies raised (Löbler and Klämbt, 1985a; Shimomura et al. 1986; Napier et al., 1988). The ABP cDNA has been cloned, sequenced (Inohara, Shimomura, Fukui, and Futai, 1989; Hesse, Feldwisch, Balschusemann, Bauw, Puype, Vandekerckhove, Löbler, Klämbt, Schell, and Palme, 1989; Tillmann, Viola, Kayser, Siemeister, Hesse, Palme, Löbler, and Klämbt, 1989), mapped (Löbler and Hirsch, 1990) and two maize genes have been analysed (Lazarus, Napier, Yu, Lynas, and Venis, 1991; Yu and Lazarus, 1991). The ABP sequence is highly conserved between maize and two dicotyledonous plants, *Arabidopsis* (S. Shimomura, *personal communication*) and strawberry (C. Lazarus, *personal communication*) with about 65% homology in each case at the amino acid level.

The sequence data provided little information about the structure or function of ABP, no homology being found to other sequences in the data bases (Hesse et al., 1989). Three features did stand out, the N-terminal signal peptide for entry into the endoplasmic reticulum (ER), the C-terminal tetrapeptide KDEL (lysine, aspartic acid, glutamic acid, leucine) which is a consensus sequence for proteins retained within the lumen of the ER (Munro and Pelham, 1987) and the mainly hydrophilic character of the protein. The only stretch of hydrophobic amino acids long enough to span a membrane as an α-helix lies in the signal peptide and this is cleaved on entry into the ER (Inohara et al., 1989). With the signal peptide cleaved, the mature ABP behaves as a soluble protein if released from the membrane fraction during purification.

The bulk of ABP appears to be located in the ER (Ray, 1977; Shimomura, Inohara, Fukui, and Futai, 1988; Jones, Lamerson, and Venis, 1989) which is consistent with the KDEL targeting sequence (Pelham, 1990). However, there is mounting evidence for ABP being functional as an auxin receptor on the outside of the plasma membrane (Barbier-Brygoo, Ephritikhine, Klämbt, Ghislain, and Guern, 1989; Barbier-Brygoo, Ephritikhine, Klämbt, Maurel, Palme, Schell, and Guern, 1991; Venis, Thomas, Barbier-Brygoo, Ephritikhine, and Guern, 1990), and this discrepancy can be addressed with the antibodies we have to ABP and other ER proteins.

We describe here some of the ways in which we are using our family of monoclonal and polyclonal antibodies to analyse ABP structure and function and its localisation within the cell.

2. Methods

Details about raising antibodies to ABP can be found in Napier et al. (1988), Hesse et al. (1989) and Löbler and Klämbt (1985a) and to the anti-ABP peptide in Venis et al. (in preparation). Descriptions of our work with the epitope mapping kit and of the immunofluorescence microscopy are also in preparation. Where we describe results from using the tobacco protoplast plasma membrane hyperpolarisation assay, the experimental protocol can be found in Barbier-Brygoo et al. (1989).

3. Results

3.1 Epitope mapping

We synthesised all the overlapping hexapeptides of ABP based on the sequence data of Tillmann *et al.* (1989). The signal peptide was not included and so peptide 1 represents the six N-terminal residues of the mature (signal peptide cleaved) protein (see Figure 1). These peptides were then used in an ELISA to identify which peptides were recognised by each antibody. The monoclonal antibody MAC 257 (Napier *et al.*, 1988) binds to peptides 8-10 (Figure 1) and so the four amino acids RDIS seem to form the principal determinant for this antibody.

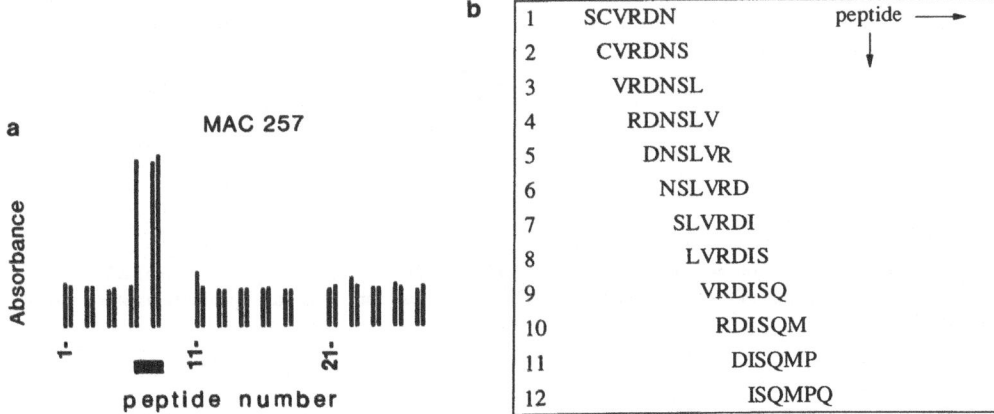

Figure 1 Epitope mapping monoclonal antibody MAC 257.
Figure 1a shows a graph of the ELISA absorbance generated to each peptide by antibody MAC 257. Peptide 1 represents the six N-terminal residues of the ABP sequence (omitting the signal peptide, see text). The amino acids are shown in Figure 1b in the single letter code. Peptide 2 represents residues 2-7 and so on for all 163 residues of the mature ABP. The elevated absorbance at peptides 8, 9 and 10 (Figure 1a), above a constant low background absorbance on all other peptides, defines the epitope for MAC 257. Only four amino acids (RDIS) occur in all three peptides.

The same peptides have been used to map polyclonal sera (Figure 2). Instead of reaction by the sera to most of the protein, only a few discrete regions of ABP are picked out. We show our own antiserum in the Figure and three dominant epitopes are identified, numbered I, II and III. Two or three of these epitopes have been found in all the anti-maize ABP sera mapped, using sera from K. Palme (Köln) and D. Klämbt (Bonn), and they are viewed, therefore, as immunodominant.

3.2 Surface epitopes

Much evidence suggests that most, perhaps all, of the surface of a protein antigen is antigenic (Laver, Air, Webster, and Smith-Gill, 1990). The ABP used as an antigen for the polyclonal sera described above was not denatured and so the epitopes we detected may all lie on the surface of the folded protein. To test this we eluted all the antibodies bound to the mapping peptides and used them in a separate ELISA against plates coated with non-denatured ABP. Only antibodies capable of recognising residues on the surface of folded ABP will be detected. When the results are plotted as a map (see Figure 2) all the epitopes recognised by our serum (Figure 2) were detected with the exception of the epitope represented by peptides 20-22 (not shown).

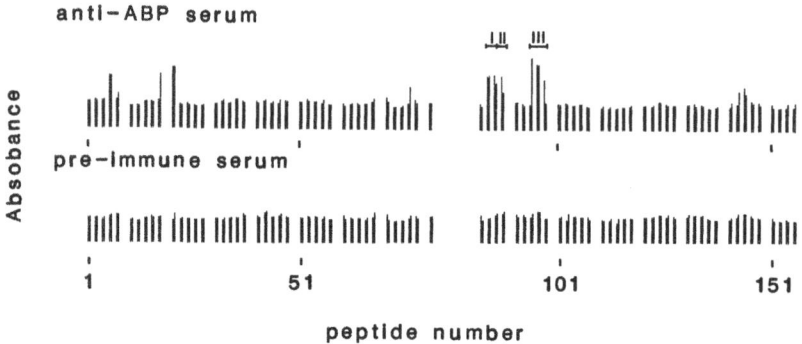

Figure 2 Epitope map of anti-maize ABP serum.
This graph shows results from the complete set of peptides. Although the whole protein was used as antigen only a few discrete regions of ABP are recognised. Three epitopes found to occur repeatedly in other antisera are labelled I-III.

3.3 Conserved epitopes

Our polyclonal serum recognises specifically ABP homologues on immunoblots in a range of plants other than maize (Venis & Napier, 1990; Napier & Venis, in prep.) including barnyard grass, mung bean, pea and rape. We have used immunoblots of barnyard grass and mung bean as immunoaffinity matrices for purifying the fraction of immunoglobulins that cross-reacts against each species. The purified immunoglobulin fraction was then mapped to identify conserved epitopes.

In barnyard grass two of the three immunodominant epitopes are conserved, epitopes I and III (Figure 3). In mung bean two proteins cross-react with the anti-maize ABP serum and these were treated separately. On the 24 kDa homologue epitope III was found to be conserved while on a 40 kDa homologue epitope I was conserved, (not shown, Napier and Venis, in preparation). In neither case were antibodies to the other immunodominant epitope, number II, detected nor were any of the other epitopes recognised by the serum.

3.4 Anti-peptide serum

A number of amino acid side group modifying agents have been shown to inhibit auxin binding (Venis, 1977; Nave and Benveniste, 1984). On the basis of these reports a stretch of the ABP sequence was considered likely to form part of the auxin binding site. A peptide embracing this region was synthesised and antisera raised to it. Of the antisera, one stains all isoforms of maize ABP and cross-reacts with homologues in a number of other plant species (Venis *et al.*, in preparation). Immunoglobulins prepared from this serum are also found to be agonists of auxin action in the tobacco protoplast hyperpolarisation assay of Barbier-Brygoo and co-workers (Venis *et al.*, in preparation). In the assay the trans-plasma membrane potential difference is measured with a microelectrode. Addition of the antibody induces a rapid hyperpolarisation characteristic of that produced by auxin (Barbier-Brygoo *et al.*, 1989, 1991).

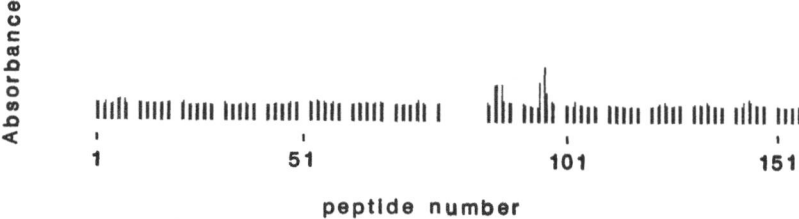

Figure 3 Conserved epitopes in barnyard grass.
Immunoblots of partially purified barnyard grass ABP were used to affinity purify those antibodies from anti-maize ABP serum which specifically cross-react. The antibodies eluted from the blots were mapped to epitope numbers I and III only.

3.5   Conformational change and ER retention

Two of the monoclonal antibodies we have raised to ABP are displaced by active auxins in a sandwich ELISA assay (Napier and Venis, 1990). We interpreted this displacement as being the consequence of an auxin-induced conformational change in ABP, the change leading to occlusion of the epitope for these antibodies. By a variety of means we have shown that these two monoclonals bind to the KDEL ER-retention terminus (Napier *et al.*, in preparation) and that a free terminal carboxylic acid group is an essential requirement for recognition. Indeed, the antibodies are extremely good markers for animal ER.

If KDEL is the single ER retention sequence in plants, as in animals (Pelham, 1990), we would expect our anti-KDEL monoclonals to stain on blots, not only of ABP, but of other ER proteins as well. We do not find this. From the literature it is becoming apparent that HDEL is a more common retention sequence in plants (Fontes, Shank, Wrobel, Moose, O'Brian, Wurtzel, and Boston, 1990; Denecke, Goldman, Demolder, Seurinck, and Botterman, 1991), as it is in some yeasts (Pelham 1990). We have access to an anti-HDEL monoclonal produced by Lewis and Pelham (MRC, Cambridge) and on blots this antibody stains a collection of proteins in plant microsomal membrane fractions, rather like the pattern obtained from animal membranes and our anti-KDEL monoclonal (Figure 4). The anti-HDEL monoclonal also stains plant ER very well at both the light and electron microscope levels (not shown). Immunofluorescence microscopy in collaboration with Dr. C. Hawes (Oxford) using anti-KDEL monoclonals shows faint staining of plant (maize) ER, but in the cortex of the cell near the plasma membrane discrete areas are intensely labelled and so is the phragmoplast during cell plate formation.

We cannot explain this particular distribution of KDEL proteins (predominantly ABP) because we have been unable to identify the KDEL-rich compartment. One possibility is the cortical ER, an ER network considered to have functions distinct from the rough ER, such as calcium storage and signal transduction (Hepler, Palevitz, Lancelle, McCauley, and Lichtscheidl, 1990). Alternatively it may represent a part of the endomembrane system through which ABP passes on the way to the plasma membrane.

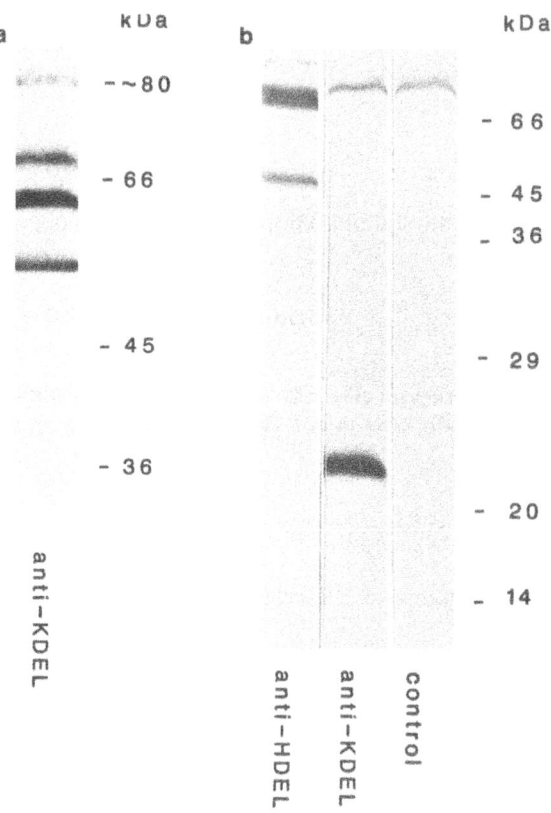

Figure 4 Immunoblotting with anti-KDEL and anti-HDEL monoclonal antibodies.
In Figure 4a mouse liver microsomal membranes have been separated by SDS-PAGE and immunoblotted with anti-KDEL monoclonal culture supernatant. In Figure 4b maize microsomal membranes have been immunoblotted with anti-HDEL, anti-KDEL and control culture supernatants. One protein stains non-specifically (see control lane). The anti-KDEL monoclonal stains ABP. The anti-HDEL monoclonal detects a number of plant proteins.

4.   Discussion

Through epitope mapping with synthetic hexapeptides we have identified immunodominant regions of ABP. Other epitopes are picked out by particular sera but only about 15% of the protein is recognised by any of the antibodies we have mapped. The three immunodominant epitopes are clustered around the single N-glycosylation site on maize ABP, which is between the epitopes numbered II and III (Figure 2). The oligosaccharide plays an important role in antigenicity in the region of ABP represented by epitope III (Napier and Venis, in preparation) and this is likely to be a reflection of the role the oligosaccharide plays in protein folding and rigidity (Wormald, Wooten, Bazzo, Edge, Feinstein, Rademacher, and Dwek, 1991).

Eluting antibodies from the epitope mapping peptides and screening for binding to intact, non-denatured ABP suggests that most of the epitopes recognised by our serum, including the immunodominant ones, lie on the surface of the folded protein. In addition, the limited number of peptides recognised, and other observations (Napier and Venis, in preparation), suggests that ABP is tightly folded.

In order to function as a receptor ABP must interact with other proteins in the signal transduction pathway. Any part of the surface may be important for such protein-protein interactions and so perhaps it is not surprising that anti-ABP sera can inhibit auxin-induced responses (Löbler and Klämbt, 1985b; Barbier-Brygoo *et al.*, 1989) if most antibodies bind to surface residues. In the experiments of Barbier-Brygoo *et al.* (1989) anti-maize ABP immunoglobulins inhibited auxin-induced membrane hyperpolarisation in tobacco protoplasts. The antibodies, which were raised to maize ABP, were presumably able to recognise conserved epitopes on a homologue of ABP on tobacco plasma membranes. Our anti-maize ABP serum also shows cross-reactivity to homologues in other plant species in immunoblotting experiments (although for technical reasons we do not get consistent results from tobacco) and we have demonstrated that two of the three immunodominant maize epitopes are conserved in barnyard grass and on two separate proteins in the dicotyledonous plant mung bean. It is possible that these conserved epitopes are those present on the surface of the tobacco homologue and that when antibodies bind the tobacco protein can no longer function as an auxin receptor. We should be able to test this hypothesis by blocking antibody inhibition of the auxin response with synthetic peptides representing these stretches of ABP sequence.

Another structural feature we have identified, at least in part, is the auxin binding site of ABP. Antisera to a synthetic peptide recognise ABP homologues in a range of plant species and can act as auxin agonists in the tobacco protoplast assay (Venis *et al.*, in preparation). In some species other proteins are recognised by these sera (Venis *et al.*, in preparation), and not by control pre-immune sera, but these have not been characterised yet.

The characteristics of both anti-peptide and anti-ABP sera have been established using the tobacco protoplast assay (Venis *et al.*, in preparation; Barbier-Brygoo *et al.*, 1989). This assay for auxin activity has consistently shown that one site of auxin perception and the site of action of ABP is on the outside of the plasma membrane (Barbier-Brygoo *et al.*, 1989, 1991; Venis *et al.*, 1990). The work of Klämbt and co-workers also suggests the same (Löbler & Klämbt 1985b; Knauth & Klämbt, 1990). Such a localisation is at odds with the known distribution of ABP in maize cells, where it has repeatedly been shown to migrate on sucrose gradients with ER marker activities (Ray, 1977; Shimomura *et al.*, 1988; Jones *et al.*, 1989). It is too early yet for us to be able to explain these discrepancies although models can be suggested. One possibility is for auxin to bind to ABP in the ER, inducing a conformational change that masks KDEL and allows ABP to escape to the cell surface. We can test this model with our antibodies, for it predicts that auxin would induce an increase of ABP secretion and the appearance of KDEL protein on the plasma membrane. A more elaborate version of this model incorporating cell wall precursors has been proposed by Cross (1991). Alternatively, minor isoforms of ABP (Hesse *et al.*, 1989) lacking KDEL may pass to the cell surface.

If auxin itself induces secretion of its own receptor then auxin is perceived in the ER as well as on the plasma membrane. There is some evidence for auxin perception being an intracellular event (Davies, 1974; Vesper and Kuss, 1990), at least for cell elongation. However, the tobacco protoplast experiments show that auxin can mediate membrane hyperpolarisation at the plasma membrane regardless of events in the ER. Clearly, the two situations are not mutually exclusive. It is possible that separate events could be mediated by the same family of proteins in different parts of the cell. It will be important to find out if ABP has a function in the ER where it is much more abundant than on the plasma membrane and for which it has a specific targeting sequence.

Acknowledgements

We would like to thank Prof. D. Klämbt and Dr. K. Palme for supplying antisera, Dr. M. Clark for help with epitope mapping software, Dr. C. Hawes and Dr. L. Fowke for immunofluorescence microscopy, Dr. H.R.B. Pelham for the anti-HDEL monoclonal antibody, Dr. C. Lazarus and Dr. S. Shimomura for communicating sequence data and Sue Tolhurst for excellent technical assistance. This work was partly supported under the BRIDGE programme of the European Economic Communities.

REFERENCES

BARBIER-BRYGOO, H., EPHRITIKHINE, G., KLAMBT, D., GHISLAIN, M., and GUERN, J., 1989. Functional evidence for an auxin receptor at the plasmalemma of tobacco mesophyll protoplasts. *Proceedings of the National Academy of Science, USA*, **86**, 891-95.

BARBIER-BRYGOO, H., EPHRITIKHINE, G., KLAMBT, D., MAUREL, C., PALME, K., SCHELL, J., and GUERN, J., 1991. Perception of the auxin signal at the plasma membrane of tobacco mesophyll protoplasts. *The Plant Journal*, **1**, 83-94.

CROSS, J.W., 1991. Cycling of auxin-binding protein through the plant cell: pathways in auxin signal transduction. *The New Biologist*, **3**, 813-19.

DAVIES, P.J., 1974. The uptake and elution of indoleacetic acid by pea stem sections in relation to auxin-induced growth. In *Plant Growth Substances 1973*. Ed Y. Sumiki, Tokyo, Hirokawa Pub. Co., 767-79.

DENECKE, J., GOLDMAN, M-H.S., DEMOLDER, J., SEURINCK, J., and BOTTERMAN, J., 1991. The tobacco luminal binding protein is encoded by a multigene family. *The Plant Cell*, **3**, 1025-1035.

FONTES, E.B.P., SHANK, B.B., WROBEL, R.L., MOOSE, S.P., O'BRIAN, G.R., WURTZEL, E.T., and BOSTON, R.S., 1991. Characterisation of an immunoglobulin binding protein homolog in the maize *floury-2* endosperm mutant. *The Plant Cell*, **3**, 483-96.

HEPLER, P.K., PALEVITZ, B.A., LANCELLE, S.A., McCAULEY, M.M., and LICHTSCHEIDL, I., 1990. Cortical endoplasmic reticulum in plants. *Journal of Cell Science*, **96**, 355-73.

HESSE, T., FELDWISCH, J., BALSCHUSEMANN, D., BAUW, G., PUYPE, M., VANDEKERCKHOVE, J., LOBLER, M., KLAMBT, D., SCHELL, J., and PALME, K., 1989. Molecular cloning and structural analysis of a gene from Zea mays (L.) coding for a putative receptor for the plant hormone auxin. *EMBO Journal*, **8**, 2453-60.

INOHARA, N., SHIMOMURA, S., FUKUI, T., and FUTAI, M., 1989. Auxin-binding protein located in the endoplasmic reticulum of maize shoots: Molecular cloning and complete primary structure. *Proceedings of the National Academy of Sciences, USA*, **86**, 3564-68.

JONES, A.M., LAMERSON, P., and VENIS, M.A., 1989. Comparison of Site I auxin binding and a 22-kilodalton auxin-binding protein in maize. *Planta*, **179**, 409-13.

KNAUTH, B., and KLAMBT, D., 1990. Is cell elongation regulated by extracellular auxin? *Botanica Acta*, **103**, 103-6.

LAVER, W.G., AIR, G.M., WEBSTER, R.G., and SMITH-GILL, S.J., 1990. Epitopes on protein antigens: Misconceptions and Realities. *Cell*, **61**, 553-56.

LAZARUS, C.M., NAPIER, R.M., YU L-X., LYNAS, C., and VENIS, M.A., 1991. Auxin binding protein - antibodies and genes. *The Molecular Biology of Plant Development*. SEB symposium Volume 45. Eds G.I. Jenkins, and W. Schuch, Company of Biologists, Cambridge.

LÖBLER, M., and HIRSCH, A.M., 1990. RFLP mapping of the abp1 locus in maize (Zea mays L.). *Plant Molecular Biology*, **15**, 513-16.

LÖBLER, M., and KLÄMBT, D., 1985a. Auxin-binding protein from coleoptile membranes of corn (Zea mays L.): 1. Purification by immunological methods and characterization. *Journal of Biological Chemistry*, **260**, 9848-53.

LÖBLER, M., and KLÄMBT, D., 1985b. Auxin-binding protein from coleoptile membranes of corn (Zea mays L.): II. Localization of a putative auxin receptor. *Journal of Biological Chemistry*, **260**, 9854-59.

MUNRO, S., and PELHAM, H.R.B., 1987. A C-terminal signal prevents secretion of luminal ER proteins. *Cell*, **48**, 899-907.

NAPIER, R.M., VENIS, M.A., BOLTON, M.A., RICHARDSON, L.I., and BUTCHER, G.W., 1988. Preparation and characterisation of monoclonal and polyclonal antibodies to maize membrane auxin-binding protein. *Planta*, **176**, 519-26.

NAPIER, R.M., and VENIS, M.A., 1990. Monoclonal antibodies detect an auxin-induced conformational change in the maize auxin-binding protein. *Planta*, **182**, 313-18.

NAVE, J.F., and BENVENISTE, P., 1984. Inactivation by phenylglyoxal of the specific binding of 1-naphthyl acetic acid with membrane-bound auxin binding sites from maize coleoptiles. *Plant Physiology*, **74**, 1035-40.

PELHAM, H.R.B., 1990. The retention signal for soluble proteins of the endoplasmic reticulum. *Trends in Biochemical Science*, **15**, 483-86.

RAY, P.M., 1977. Auxin-binding sites of maize coleoptiles are localized on membranes of the endoplasmic reticulum. *Plant Physiology*, **59**, 594-99.

SHIMOMURA, S., SOTOBAYASHI, T., FUTAI, M., and FUKUI, T., 1986. Purification and properties of an auxin-binding protein from maize shoot membranes. *Journal of Biochemistry*, **99**, 1513-24.

SHIMOMURA, S., INOHARA, N., FUKUI, T., and FUTAI, M., 1988. Different properties of two types of auxin-binding sites in membranes from maize coleoptiles. *Planta*, **175**, 558-66.

TILLMANN, U., VIOLA, G., KAYSER, B., SIEMEISTER, G., HESSE, T., PALME, K., LOBLER, M., and KLAMBT, D., 1989. cDNA clones of the auxin-binding protein from corn coleoptiles (Zea mays L.): Isolation and characterisation by immunological methods. *EMBO Journal*, **8**, 2463-67.

VENIS, M.A., 1977. Affinity labels for auxin binding sites in corn coleoptile membranes. *Planta*, **134**, 145-49.

VENIS, M.A., and NAPIER, R.M., 1990. Characterisation of auxin receptors. In *Hormone Perception and Signal Transduction in Animals and Plants*. SEB Symposium, Volume 44. Eds J. Roberts, C. Kirk and M. Venis. Company of Biologists, Cambridge. pp 55-65.

VENIS, M.A., THOMAS, E.W., BARBIER-BRYGOO, H., EPHRITIKHINE, G., and GUERN, J., 1990. Impermeant auxin analogues have auxin activity. *Planta*, **182**, 232-35.

VESPER, M.J., KUSS, C.L., 1990. Physiological evidence that the primary site of auxin action in maize coleoptiles is an intracellular site. *Planta*, **182**, 486-91.

WORMALD, M.R., WOOTEN, E.W., BAZZO, R., EDGE, C.J., FEINSTEIN, A., RADEMACHER, T.W., and DWEK, R.A., 1991. The conformational effects of N-glycosylation on the tailpiece from serum IgM. *European Journal of Biochemistry*, **198**, 131-39.

YU, L-X., and LAZARUS, C.M., 1991. Structure and sequence of an auxin-binding protein gene from maize (Zea mays L.). *Plant Molecular Biology*, **16**, 925-30.

Membrane Receptors for Fusicoccin:
A Molecular Study

**P. Aducci[1], A. Ballio[2], V. Fogliano[2], M.R. Fullone[2], M. Marra[1]
and D. Verzili[3]**

1.   Department of Biology
     II University of Rome "Tor Vergata"
     v.O.Raimondo
     00173 Rome, Italy

2.   Department of Biochemical Sciences "A. Rossi-Fanelli"
     University of Rome "La Sapienza"
     p.le A. Moro 5
     00185 Rome, Italy

3.   CNR, Centre of Molecular Biology
     Department of Biochemical Sciences "A. Rossi-Fanelli"
     University of Rome "La Sapienza"
     p.le A. Moro 5, 00185 Rome, Italy

Abbreviations: fusicoccin, FC; p-hydroxymercuribenzoic acid, PMB;
phenylmethylsulfonyl fluoride, PMSF.

1.   Introduction

The phytotoxic metabolite of the fungus *Fusicoccum amygdali* Del. (Ballio, Chain, De Leo,
Erlanger, Mauri, Tonolo, 1964) interacts with higher plant cells promoting cell extension
growth and membrane transport processes (Marrè, 1979).   The interest of plant
physiologists in this toxin has been increased by the finding that these effects are similar
to those regulated by some plant hormones.   It is now well established that all the
hormone-like responses stimulated by fusicoccin (FC) depend on the activation of a proton-
translocating plasmalemma ATPase, which is present in all higher plants and is necessary

Transport and Receptor Proteins of Plant Membranes
Edited by D.T. Cooke and D.T. Clarkson, Plenum Press, New York, 1992

for nutrient transport. The stimulation of this enzyme represents the more direct FC action, but so far its mechanism is unclear. Many studies directed to elucidate the nature of the signal triggering the H^+-ATPase showed that the specificity of FC interaction with plant cells depends on the presence of FC binding sites at the plasmalemma. Results obtained so far show that they can be considered true receptors (Aducci, Marra, Ballio, 1990). Their presence in higher plants which are not hosts of the fungus *F. amygdali* and therefore do not meet the toxin in natural conditions is explained by the occurrence, in several plants, of a physiological ligand mimicked by FC (Aducci, Crosetti, Federico, Ballio, 1980; Ballio and Aducci, 1987; Marra, Ballio, Aducci, 1988). It is very likely that the interaction of this ligand with FC receptors is the first step in a chain of reactions leading to the biological responses known to be promoted by FC. Therefore, the system recognised by FC can, under physiological conditions, play a crucial role in cell metabolism.

The evidence that all higher plants so far investigated are physiologically responsive to FC is paralleled by a widespread distribution of FC receptors. The study of these receptors was made possible by the preparation of suitable radioactive ligands (Ballio, Federico, Pessi, Scalorbi, 1980; Feyerabend and Weiler, 1988) and, more recently, by a photo-affinity radio-ligand (Feyerabend and Weiler, 1989). Their localisation and biochemical properties, in a number of plants, have been investigated by several groups (see for references Aducci, Ballio, Federico, Montesano, 1982b; Aducci and Ballio, 1987; Aducci and Ballio, 1989; Weiler, Meyer, Oecking, Feyerabend, Mithöfer, 1990).

2.   Purification of FC Receptors

Attempts to purify these proteins started after methods for their solubilisation had been developed. The crude solubilised receptors prepared by different procedures were revealed by gel filtration to have a relative molecular mass of 80-100 kDa (Pesci, Tognoli, Beffagna, Marrè, 1979; Stout and Cleland, 1980; Aducci, Ballio, Federico, 1982a; Feyerabend and Weiler, 1989).

Despite the different approaches to purification which have been attempted so far, the molecular analysis of the pure receptors has not yet been achieved. The SDS-PAGE analysis of FC-binding protein complexes from *Vicia faba* L. leaves purified by FPLC (Feyerabend and Weiler, 1989) gave a major band of 34 kDa, a result confirmed by photo-affinity labelling experiments. A similar result was obtained by partial purification of receptors from oat roots by affinity chromatography (de Boer, Watson, and Cleland, 1988). The SDS-PAGE of the fraction eluted from the affinity column showed two major protein bands with molecular weights of 29.7 kDa and 31 kDa, although a higher molecular weight protein of 67 kDa has been occasionally observed (Feyerabend and Weiler, 1989; de Boer *et al.*, 1988). Moreover, from the photo-affinity labelling studies it appears that a radio-labelled material was also present on the migrating front of SDS-PAGE gels, suggesting a non-specific interaction of FC with low molecular weight components (Feyerabend and Weiler, 1989). In a more recent paper (Oecking and Weiler, 1991) it was shown that purified fractions from plasma membranes of *Commelina communis* L. contained two polypeptides of apparent molecular masses of 30.5 and 31.6 kDa. The separation of the two polypeptides was unsuccessful even in the purest FC receptor preparations. Larger amounts of pure proteins are necessary for a more detailed analysis at the molecular level of FC receptors.

Our group has recently attempted the purification of FC receptors solubilised from maize tissues. Unlike other studies, we achieved solubilisation from acetone-dried microsomal fractions without the use of detergents (Aducci, Fullone, Ballio, 1989). The purification scheme used is reported in Table 1, where two alternative procedures are shown. The first step is represented by a conventional adsorption chromatography on hydroxyapatite (HPHT), while all others are performed by HPLC. Procedure 1, which has a gel permeation step instead of chromatofocusing, gave more reproducible results and was, therefore, routinely adopted. When the receptors were solubilised from purified two-phase partitioned plasmalemma preparations, the final purification was about 5000-fold, as shown in Table 2. Although the binding ability of the fractions was maintained throughout the purification procedure, it was better preserved when FC was bound to its receptors. Therefore, in routine work the receptors were detected by measuring the radioactivity of fractions and analysed by disc-gel electrophoresis. As with other studies on different plants, the SDS-PAGE of the purified fractions showed a molecular weight doublet in the range of 30 kDa, namely two protein bands of 31 and 33 kDa. Moreover, a higher molecular weight band at 90 kDa was also detected in the purified fractions, suggesting that the receptor may be present in a multimeric form. However, the possibility that only one of the polypeptides corresponds to the binding protein cannot be ruled out. At the moment it is unknown whether the 30 and 90 Kda bands are structurally related. A possible proteolytic degradation of FC receptors might explain the heterogeneity observed in SDS-PAGE. Therefore, we have checked the effect of p-hydroxymercuribenzoic acid (PMB) and phenylmethylsulfonyl fluoride (PMSF), specific inhibitors for serine and cysteine proteinases, respectively, on the binding activity of FC receptors at two different temperatures (Table 3). The addition of these inhibitors, effective in preserving FC binding activity at low temperature, was routinely used during all the purification steps. Their presence did not affect the final protein pattern, suggesting that the low molecular weight components were not derived from proteolytic cleavage.

Radiation-inactivation studies have suggested that FC receptors may be part of a larger complex made of different proteins involved in the transduction of the FC signal (De Michelis, Pugliarello, Olivari, Rasi-Caldogno, 1989). However, the identification of a specific element of the transduction chain is still to be achieved.

The identity of the receptor structure is necessary for the interpretation of the first events in the FC mode of action and of their biochemical regulation at the molecular level. Moreover, the determination of the primary structure of this protein may elucidate its topology within the plasma membrane and promote further studies such as gene cloning.

3.   Strategies for Signal Transduction Studies

The knowledge of the FC mode of action not only requires the structural identification of FC receptors, but also the elucidation of events that follow the primary interaction of FC with its receptors and lead to the activation of the H^+-ATPase. In order to clarify the relationship between receptors and ATPase, the analysis of natural membranes, where the enzyme stimulation was first demonstrated (Rasi-Caldogno and Pugliarello, 1985), was substituted with that of a reconstituted system obtained with isolated components of the plasma membrane, such as solubilised fractions from maize tissues containing FC receptors and H$^+$-ATPase (Aducci *et al.*, 1988; Aducci *et al.*, 1991). In this system it was shown that the binding reaction is a crucial step for the activation of the H$^+$-ATPase. This approach might be useful for the identification of transducing elements.

Table 1 Purification step sequences

	Procedure 1	Procedure 2
1	Adsorption (Bio-Gel HPHT)	Adsorption (Bio-Gel HPHT)
2	Anion exchange (Bio-Gel TSK-5-PW)	Anion exchange (Bio-Gel TSK-5-PW)
3	Gel permeation (Superose 12)	Chromatofocusing (Bio-Gel TSK-5-PW)
4	Anion exchange (Bio-Gel TSK-5-PW)	Anion exchange (Bio-Gel TSK-5-PW)

Table 2 Purification of FC receptors* from maize shoots

Chromatography step	Total protein (mg)	^{3}[H]-FC bound (DPM x 1000)	Specific activity (DPM / mg protein)	Purification fold
none	30	3000	100	-
Bio-Gel HPHT	1	1000	1000	10
Bio-Gel TSK-5-PW	0.1	420	4200	42
Superose 12	0.03	300	10,000	100
Bio-Gel TSK-5-PW	0.0004	200	500,000	5000

* Plasma membrane vesicles were purified according to Larsson, Kjellbom, Widell, Lundborg (1984). Solubilisation and tests for FC binding activity were performed according to Aducci *et al.*, (1989).

Table 3 Effect of proteinase inhibitors on the binding activity of FC receptors solubilised from maize.

Addition of inhibitors** at different pre-incubation temperature (°C)	Binding activity* (%) at different pre-incubation times (h)		
	1	24	48
27° C none	88	18	13
27° C + PMB and PMSF	86	43	16
4° C none	92	90	49
4° C + PMB and PMSF	100	99	60

*   Binding activity is expressed as the percentage of maximum.
       The binding test was performed according to Aducci *et al.*, 1989.

**   The proteinase inhibitors PMB and PMSF were added at 0.1 mM final concentration.

Electrophysiological studies (Blatt, 1988; Blatt and Clint, 1989) suggest that more than one physiological target is involved in FC action. In fact, FC could inactivate K^+ channels, as shown by its inhibition of leak current. The more complex model that can be derived from these results might receive support from the elucidation of the individual transduction steps from FC binding to the suggested plasmalemma transport system(s) possibly by designing different methodological approaches.

Acknowledgement

This research has been supported by the National Research Council of Italy (CNR), Special Project RAISA, Sub-project n.2 Paper N.254.

REFERENCES

ADUCCI, P., CROSETTI, G., FEDERICO, R., AND BALLIO, A., 1980. Fusicoccin receptors. Evidence for endogenous ligands. *Planta*, **148**, 208-210.

ADUCCI, P., BALLIO, A., AND FEDERICO, R., 1982a. Solubilisation of fusicoccin binding sites. In *Plasmalemma and tonoplast: their function in the plant cell*. Eds D. Marmè, E. Marrè, and R. Hertel. pp 279-284. Elsevier, Amsterdam.

ADUCCI, P., BALLIO, A., FEDERICO, R., and MONTESANO, L., 1982b. Studies on fusicoccin-binding sites. In *Plant growth substances*. Ed P.F. Wareing. pp 395-404. Academic Press, London.

ADUCCI, P., and BALLIO, A., 1987. The interaction of fusicoccin with specific binding sites. In *Plant hormone receptors*. Ed D. Klämbt. pp 131-139. Springer-Verlag, Berlin.

ADUCCI, P., BALLIO, A., BLEIN, J-P., FULLONE, M.R., ROSSIGNOL, M., and SCALLA, R., 1988. Functional reconstitution of a proton-translocating system responsive to fusicoccin. *Proceedings of the National Academy of Sciences, USA*, **85**, 7849-7851.

ADUCCI, P., and BALLIO, A., 1989. Mode of action of fusicoccin: the role of specific receptors. In *Phytotoxins and plant pathogenesis*. Eds A. Graniti *et al.* pp 143-150. Springer-Verlag, Berlin.

ADUCCI, P., FULLONE, M.R., and BALLIO, A., 1989. Properties of proteoliposomes containing fusicoccin receptors from maize. *Plant Physiology*, **92**, 1402-1406.

ADUCCI, P., MARRA, M., and BALLIO, A., 1990. Fusicoccin: receptors and endogenous ligands. In *Hormone perception and signal transduction in animals and plants*. Eds J. Roberts, C. Kirk, and M.A. Venis. pp 111-117. The Company of Biologists Limited, Cambridge.

ADUCCI, P., BALLIO, A., FULLONE, M.R., and MARRA, M., 1991. Biochemical characterisation of a dual reconstituted system responsive to fusicoccin. In *Proceedings of the 14th International Conference on Plant Growth Substances*, in press.

BALLIO, A., CHAIN, E.B., DE LEO, P., ERLANGER, B.F., MAURI, M. and TONOLO, A., 1964. Fusicoccin, a new wilting toxin produced by *Fusicoccum amygdali* Del. *Nature*, **203**, 296.

BALLIO, A., FEDERICO, R., PESSI, A., and SCALORBI, D., 1980. Fusicoccin binding sites in subcellular preparations of spinach leaves. *Plant Science Letters*, **18**, 39-44.

BALLIO, A., and ADUCCI, P., 1987. Search for endogenous ligands to fusicoccin binding sites. In *Plant hormone receptors*. Ed D. Klämbt. pp 125-130. Springer-Verlag, Berlin.

BLATT, M.R., 1988. Mechanism of fusicoccin action: a dominant role for secondary transport in higher plant cell. *Planta*, **174**, 187-200.

BLATT, M.R. and CLINT, G.M., 1989. Fusicoccin activates energy-coupled K^+ uptake by stomatal guard cells. *Plant Physiology*, **89**, Supplement, Abstract No. 1169, 195.

DE BOER, A.H., WATSON, B.A., and CLELAND, R.E., 1989. Purification and identification of the fusicoccin binding protein from oat root plasma membrane. *Plant Physiology*, **89**, 250-259.

DE MICHELIS, M.I., PUGLIARELLO, M.C., OLIVARI, C., and RASI-CALDOGNO, F., 1989. On the mechanism of FC-induced activation of the plasma membrane H^+-ATPase. In *Plant membrane transport: the current position*. Eds J. Dainty, M.I. De Michelis, E. Marrè, and F. Rasi-Caldogno. pp 373-378. Elsevier, Amsterdam.

FEYERABEND, M., and WEILER, E.W., 1988. Characterisation and localisation of fusicoccin-binding sites in leaf tissues of *Vicia faba* L. probed with a novel radio-ligand. *Planta*, **174**, 115-122.

FEYERABEND, M., and WEILER, E.W., 1989. Photo-affinity labelling and partial purification of the putative plant receptor for the fungal wilt-inducing toxin, fusicoccin. *Planta*, **178**, 282-290.

LARSSON, C., KJELLBOM, P., WIDELL, S., and LUNDBORG, T., 1984. Sidedness of plant plasma membrane vesicles purified by partition in aqueous two-phase systems. *FEBS Letters*, **171** 271-276.

MARRA, M., BALLIO, A., and ADUCCI, P., 1988. Immuno-affinity chromatography of fusicoccin. *Journal of Chromatography*, **440**, 47-51.

MARRÈ, E., 1979. Fusicoccin: a tool in plant physiology. *Annual Review of Plant Physiology*, **30**, 273-288.

OECKING, C., and WEILER, E.W., 1991. Characterisation and purification of the fusicoccin-binding complex from plasma membrane of *Commelina communis*. *European Journal of Biochemistry*, **199**, 685-689.

PESCI, P., TOGNOLI, L., BEFFAGNA, N., and MARRÈ, E., 1979. Solubilisation and partial purification of a fusicoccin-receptor complex from maize coleoptiles. *Plant Science Letters*, **15**, 313-317.

RASI-CALDOGNO, F., and PUGLIARELLO, M.C., 1985. Fusicoccin stimulates the H^+-ATPase of plasmalemma in isolated membrane vesicles from radish. *Biochemical and Biophysical Research Communications*, **133**, 280-285.

STOUT, R.G., and CLELAND, R.E., 1980. Partial characterisation of fusicoccin binding to receptor sites on oat root membranes. *Plant Physiology*, **66**, 353-359.

WEILER, E.W., MEYER, C., OECKING, C., FEYERABEND, M., and MITHÖFER, A., 1990. The fusicoccin receptor of higher plants. In *Plant gene transfer*. Eds C. Lamb and R. Beachy. pp 153-164. Alan R. Liss, Inc., New York.

GTP-Binding Proteins in Higher Plant Cells

I.R. White, A. Wise, P.M. Finan, J. Clarkson and P.A. Millner[1]

Department of Biochemistry and Molecular Biology
University of Leeds
Leeds, LS2 9JT, UK

1.   To whom correspondence should be addressed.

Keywords: G-protein, plant cell, GTP-binding, antipeptide antibodies.

Abbreviations: G-protein, GTP-binding protein; GTPγS, guanosine-5´-0-(3-thiotriphosphate); MEGA-9, nonanoyl-N-methylglucamide; CHAPS, 3-[(cholamidopropyl)-dimethylammonio]-1-propane-sulphonate.

1.   Introduction

Molecular interactions driven by GTP binding and its subsequent hydrolysis have emerged in such diverse areas as protein synthesis, polymerisation/depolymerisation of cytoskeletal proteins and transduction of signals which are perceived initially by cell surface located receptors (Bourne, Sanders, and McCormick, 1990).   Within the latter category of interactions, which are effected by the signal transducing G-proteins two super-families are evident. These are the monomeric or small G-proteins, typified by Ras and related proteins (Barbacid, 1987;   Burgoyne, 1989) and the heterotrimeric G-proteins (Birnbaumer, Abramowitz, and Brown, 1990; Kaziro, Itoh, Kozasa, Nakafuku, and Satoh, 1991). Whilst members of the former family have undoubtedly been identified in plant cells (Drobak, Allan, Comerford, Roberts, and Dawson, 1988) this report will mainly concern itself with evidence for, and progress in, the isolation of members of the heterotrimeric G-protein family.

To date, most evidence for the presence of plant G-proteins has relied on two or three lines of circumstantial evidence. Non-hydrolysable GTP analogues, such as GTPγS were shown to stimulate the turnover of inositol phospholipids (Dillenschneider, Hetherington, Graziana, Alibert, Berta, Haiech, and Ranjeva, 1986), whilst a number of workers have measured specific high affinity binding of GTP or GTPγS to thylakoids (Millner, 1987),

Transport and Receptor Proteins of Plant Membranes
Edited by D.T. Cooke and D.T. Clarkson, Plenum Press, New York, 1992

185

microsomal membrane fractions (Jacobs, Thelen, Farndale, Astle, and Rubery, 1988; Hasunuma and Funadera, 1988; Hasunuma, Furukawa, Tonita, Mukai, and Nakamura, 1988; Zbell, Schwendemann and Bopp, 1989) and highly resolved plasmalemma (Blum, Hinsch, Schulz, and Weiler, 1988; Wise and Millner, 1991) from a number of plant species. Generally, binding constants of the order of 10^{-8} to 10^{-7} M were reported in these studies, similar to those reported for G-proteins from animal cells and lower eukaryotes such as yeast.

Much use has also been made of the highly conserved nature of the G$\propto$ and Gβ subunits sequences. Peptides, corresponding in sequence to regions within animal G-protein subunits (Mumby, Kahn, Manning, and Gilman, 1985; Goldsmith, Giersclick, Milligan, Unson, Vinitsky, Malech, and Spiegel, 1987) were used as synthetic antigens to prepare antibodies, directed to conserved regions for G-protein subunits, i.e. the "$\propto_{common}$" sequence, or specific for G-protein subclasses, e.g. G$\propto_i$. Such antisera have also been used to identify putative G$\propto$-subunits associated with plant cell membranes (Blum *et al.*, 1988; Jacobs *et al.*, 1988; Clarkson, Finan, Ricart, White, and Millner, 1990). Finally, using a molecular biological approach, genes encoding G$\propto$ and Gβ homologues (Ma, Yanofsky, and Meyerowitz, 1990; Schloss, 1990) were cloned from *Arabidopsis thaliana* and *Chlamydomonas rheinhardtii* respectively. The protein encoded by the *Arabidopsis* gene was of 44 kDa molecular mass and showed substantial similarity to the Gi subgroup of G-proteins.

Although work so far provides a strong indication of heterotrimeric signal-transducing G-proteins in plant cells, so far these have not been isolated. In the present contribution, we describe the immunological identification of G-protein subtypes from etiolated *Pisum sativum* seedlings and from *A. thaliana* leaf tissue. In addition we describe the partial purification of G-proteins from these sources.

2. Materials and Methods

2.1 Immunological procedures

Synthetic peptides, corresponding to sequences within animal G-proteins (see Table 1) were prepared by solid phase synthesis (Atherton and Shepherd, 1990) and conjugated to PPD prior to immunisation of rabbits (Lachmann, Strangeways, Vyakarnam, and Evan, 1986). Antisera were used without further preparation or after isolation of the IgG fraction using protein-G Sepharose.

Table 1 Dissociation constant (Kd) and number of sites (n) for binding of GTPγS to *Zea mays* and *Nicotiana tabacum* plasmalemma. Values of Kd and n were determined by Scatchard analysis of binding data.

Species	Kd (nM)	n (pmol mg^{-1}protein)
Nicotiana Tabacum	26	120
Zea mays	27	40

2.2 G-Protein isolation procedures

Microsomal membranes were prepared from etiolated *P.sativum* seedlings or from *A.thaliana* leaves according to Kjellbom and Larsson (1984) and resuspended in medium TEDM, which comprised 25 mM Tris-HCl, pH 8.0, 1 mM EDTA, 1 mM DTT and 5 mM $Mg(CH_3COO)_2$. The *Arabidopsis* microsomal membranes were then treated with 1 M NaCl or 1 mM EDTA for 1 hour prior to centrifugation at 40,000 g for 1 hour. *Pisum* membranes were treated with various detergents in TEDM (see Table 2) for 30 minutes at 4° C prior to centrifugation at 288,000 xg for 30 minutes. In both cases, the post-centrifugation supernatants were analysed for [^{35}S]GTPγS binding and following separation of polypeptides by SDS-PAGE, for immunologically detectable G$\propto$ subunits.

2.3 [^{35}S]GTPγS binding

Binding was assayed, at 30° C in a medium comprising 20 mM tricine-NaOH, pH 7.6, 100 mM NaCl, 1 mM EDTA, 1 mM DTT, 30 mM $Mg(CH_3COO)_2$, 2-500 nM of unlabelled GTPγS, 9.2 kBq of [^{35}S]GTPγS and approximately 20 μg of membranes or solubilised protein. Non-specific binding was measured by inclusion of 20 μM GTP in the assay mixture. Following incubation for 10 minutes to 1 hour, the protein was collected by filtration on to nitrocellulose discs (0.2 μm pore) and rapidly washed with 4 x 2 ml of ice cold 20 mM tricine-NaOH, pH 7.6, 100 mM NaCl, 25 mM $Mg(CH_3COO)_2$. Discs were air dried and then solubilised in 2 ml of ethoxyethanol prior to determination of [^{35}S]-GTPγS by scintillation counting.

2.4 Chromatography of *P. sativum* polypeptides

Proteins released from about 50 mg of *P. sativum* microsomal membranes by hypotonic washing in TEDM, or by incubation with 25 mM MEGA-9, were applied to a Pharmacia MONO Q anion exchange column in TEDM, pH 8.3 and eluted with a 0 to 1 M gradient of NaCl in the same medium. For proteins released by MEGA-9, 10 mM of this detergent was also present. The column eluate was monitored for absorbance at 280 nm and the fractions collected assayed for [^{35}S]GTPγS binding.

3. Results and Discussion

In common with other laboratories we have measured high affinity [^{35}S]GTPγS binding to purified plasmalemma derived from a number of species, including the dicots *Nicotiana tabacum*, *P.sativum* and *A.thaliana* and the monocots *Zea mays* and *Triticum aestivum*. Representative values for the binding constant, Kd, and the number of sites, n, are shown in Table 1. The Scatchard plots, from which these values were derived, indicated that a single class of binding site was present, although it would be difficult to distinguish between several sites if they were all of a similar Kd. The values of Kd shown were in the range reported by other laboratories (Hasunuma and Funadera, 1988; Jacobs *et al.*, 1988, Blum *et al.*, 1988) and are consistent with those found for animal G-proteins (Birnbaumer, Abramowitz and Brown, 1990). Antisera raised against synthetic peptides which correspond in sequence to regions within animal G$\propto$ subunits and within the *Arabidopsis* G$\propto$, GPA1, have also proved successful in identifying candidate G-proteins. Figure 1 shows that the Gi$_{common}$ antibody (Figure 1A) directed towards the C-terminus of the Gi$\propto$ subtype, cross-reacted with M_r 37

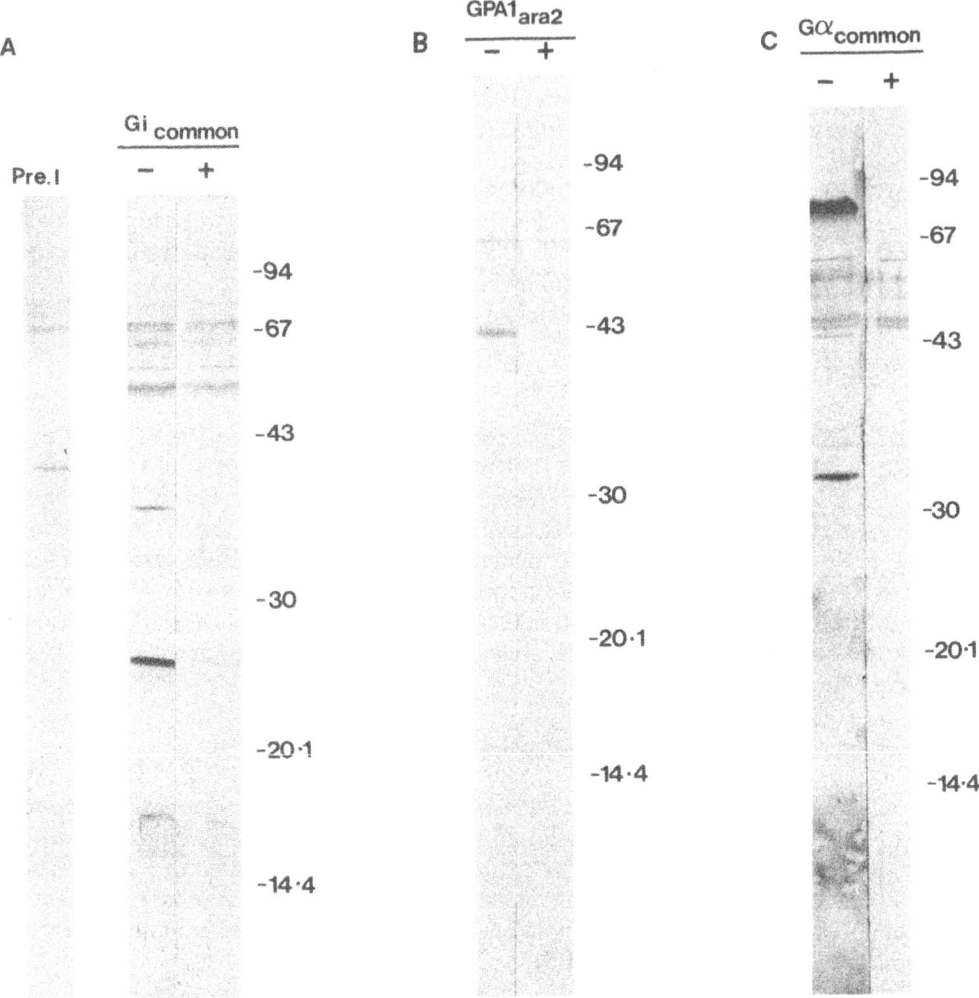

Figure 1 Immunological identification of Gα subtypes present in *P. sativum* microsomal membranes. Membranes equal to 40 μg of protein were electrophoresed on 12% SDS-polyacrylamide gels prior to transfer to Immobilon-P membrane and probing with the antisera indicated. (A), Gi$_{common}$; (B), GPA1$_{ara2}$; (C), Gα$_{common}$; -/+ antisera preincubated in the absence or presence of 33 μg ml^{-1} of the appropriate synthetic peptide. Pre I, pre-immune serum.

kDa and 25 kDa proteins in *P. sativum* membranes, whilst the GPA1$_{ara2}$ antibody (Figure 1B), directed towards the equivalent sequence in GPA1, detected a 43 kDa polypeptide in these membranes. Use of the Gα$_{common}$ antibody which has also been employed by others (Blum *et al.*, 1988;) indicated a 33 kDa cross-reacting protein (Figure 1C). In similar experiments with *A. thaliana* membranes, the Gi$_{common}$ antibody cross-reacted with a M$_r$7 kDa protein whilst the GPA1$_{ara2}$ antibody identified M$_r$ 43 kDa and 37 kDa polypeptides. An important control in these experiments (Figure 1A,B,C) is the removal of cross-reactivity when the antisera were preincubated with the synthetic peptides to which they were raised. In some reports, the lack of this control has led to the identification of G-proteins which certainly represented non-specifically cross-reacting material.

Attempts to purify G-proteins were initiated by determining the conditions required for release of GTPγS binding activity and of immunologically identified proteins from these membranes. In Table 2, various treatments aimed at the removal of functional G-proteins from *P. sativum* membranes, hypotonic washing procedures and treatment with a number of detergents were effective at releasing [^{35}S]GTPγS binding. All of the detergents appeared to interfere with binding to some extent, with β-octylglucoside and MEGA-9 providing the best retention of binding activity. When blots of the proteins released were probed with the GPA1$_{ara2}$ antibody (not shown), the 43 kDa protein (see Figure 1B) was only present in the detergent-derived material, indicating that another G-protein must have been responsible for the [^{35}S]GTPγS binding activity released by the hypotonic washing procedure.

Table 2 Release of high affinity GTPγS binding from *P. sativum* and *A. thaliana* microsomal membranes. Membranes were subjected to the treatments indicated, prior to recovery of proteins released by sedimentation of the microsomal membranes (see "Materials and Methods"). Binding of [^{35}S]-GTPγS to the proteins released was monitored using about 20 μg protein, 9.2 KBq [^{35}S]-GTPγS and either 5 nM (*P. sativum* proteins) or 50 nM (*A. thaliana* proteins) unlabelled GTPγS. Control membranes represent membranes that were untreated.

Species	Treatment		GTPγS binding Specific activity (pmol mg^{-1})
Pisum	1%(w/v)	Na-cholate	0.53
sativum	15 mM	CHAPS	0.57
	25 mM	β-octylglucoside	0.87
	25 mM	MEGA-9	0.97
	0.5%(v/v)	Lubrol	0.33
	0.5%(v/v)	Triton X-100	0.23
	TEDM	(no detergent)	2.67
	Control	Membranes	1.66
Arabidopsis	1 M	NaCl	6.07
thaliana	1 mM	EDTA	8.00
	Control	Membranes	8.71

With *A. thaliana* microsomal membranes, treatment with high concentrations of NaCl seemed to be the most effective method for releasing [^{35}S]GTPγS binding, although the material released had a lower specific binding activity than that removed by EDTA. The NaCl treatment appeared to release a 37 kDa Gi$_{common}$ reactive polypeptide, but not the 43 kDa protein which cross-reacted with antibody GPA1$_{ara2}$ (not shown). It should be noted that the values in Table 2 are only relative. Since the measurements on *P. sativum* proteins were performed at a concentration below Kd, the values presented must represent an underestimate of the true specific activity.

Subsequent separation of the proteins released by hypotonic washing with TEDM medium, or with the neutral detergent MEGA-9, by anion exchange chromatography, yielded a small number of peaks of [^{35}S]GTPγS binding activity which were well resolved from the bulk of the material applied to the column. In both cases, a peak of activity was eluted at

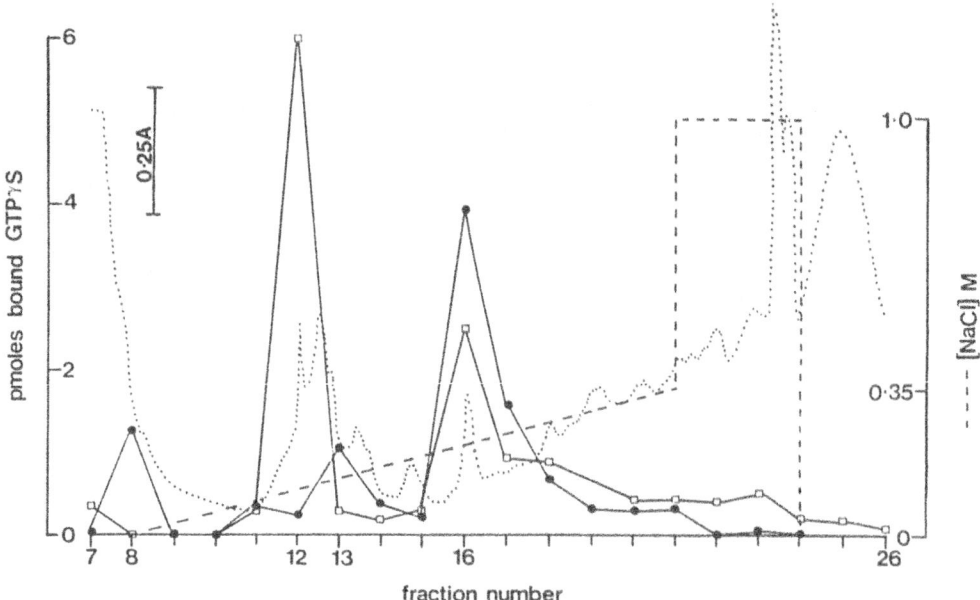

Table 3. This discovery is of particular significance since members of this family of proteins may be involved in G-protein function (Kikkawa, Takahashi, Takahashi, Simada, Ui, Kimura, and Katada, 1990).

Clearly, from the data presented, we have shown that a number of G-protein subtypes are present within plant cells, including those which are homologous to Gi∝ and to GPA1, and those which carry the G∝_common sequence. In addition, as phosphatidylinositol lipid signalling is emerging as an important regulatory pathway in plant cells (Einspahr and Thompson, 1990), it is highly likely that the Gq sub-family of G-proteins which modulate phospholipase C activity will also be found.

Table 3 Alignment of *P. sativum* nucleoside diphosphate kinase NDK-P1 and NDK-P2 partial sequences with other nucleoside diphosphate sequences.   NM23-H1 and NM23-H2, metastasis suppressor proteins; AWD$DROME, *Drosophila* abnormal wing disc protein; NDK$MYXXA; *Myxococcus xanthus* nucleoside diphosphate kinase; DDINDK, *Dictyostelium discoideum* nucleoside diphosphate kinase.

```
NDK1$Pea      -----------FIAIKPDGVQRGLVGEIISRFEKKGFYLKGLKFVN--
NDK2$Pea      -----------FIAIKPDGVQRGLVSEIISRFEKKGFYLKGLKFVN--


NM23-H1       MAN-----CERTFIAIKPDGVQRGLVGEIIKRFEQKGFRLVGLKFMQA-
NM23-H2       MAN-----LERTFIAIKPDGVQRGLVGEIIKRFEQKGFRLVAMKFLRA-
AWD$Drome     MAA----NKERTFIMVKPDGVQRGLVGKIIERFEQKGFKLVALKFTWA-
DDINDK        MSTNKVNKERTFLAVKPDGVARGLVGEIIARYEKKGFVLVGLKQLVP-
NDK$mYXXA     MAI------ERTLSIIKPDGLEKGVIGKIISRFEEKGLKPVAIR-LQHL


NDK1$Pea      VERLLEKHYADLSAKPFFSGLVDYIISGPVVAMIWEGKNV
NDK2$Pea      VERLLEKHYADLSAKPSFSGLVDYIISGPVVAMIWEG

NM230H1       SEDLLKEHYVDLKDRPFFAGLVKYMHSGPVVAMVWEGLNVV
NM23-H2       SEEHLKQHYIDLKDRPFFPGLVKYMNSGPVVAMVWEGLNVV
AWD$Drome     SKELLEKHYADLSARPFFPGLVNYMNSGPVVPMVWEGLNVV
DDINDK        TKDLAESHYAEHKERPFFGGLVSFITSGPVVAMVFEGKGVV
NDK$Myxxa     SQAQAEGFYAVHKARPFFKDLVQFMISGPVVLMVLEGENAV
```

REFERENCES

ATHERTON, E., and SHEPHARD, R.C., 1990. Solid phase peptide synthesis. IRL Press, Oxford.

BARBACID, M. 1987. *ras* Genes. *Annual Review of Biochemistry*, **56**, 779-827.

BIRNBAUMER, L., ABRAMOWITZ, J., and BROWN, A.M., 1990. Receptor-effector coupling by G-Proteins. *Biochimica et Biophysica Acta*, **1031**, 163-224.

BLUM, W., HINSCH, K.-D., SCHULZ, G., and WEILER, E.W., 1988. Identification of G-proteins in the plasma membranes of higher plants. *Biochemical and Biophysical Research Communication*, **156**, 954-959.

BOURNE, H.R., SANDERS, D.A., and McCORMICK, F., 1990. The GTPase superfamily: a conserved switch for diverse cell functions. *Nature*, **348**, 125-132.

BURGOYNE, R.D., 1989. Small GTP-binding proteins. *TIBS 14*, 394-396.

CLARKSON, J., FINAN, P.M., RICART, C.A., WHITE, I.R., and MILLNER, P.A., 1991. Specific immunodetection of G-protein in plant cell membranes. *Biochemical Society Transactions*, **19**, 239S.

DILLENSCHNEIDER, M., HETHERINGTON, A., GRAZIANA, A., ALIBERT, G., BERTA, T., HAIECH, J., and RANJEVA, R., 1986. The formation of inositol phosphate derivatives by isolated membranes from *Acer pseudoplatinus* is stimulated by guanine nucleotides. *FEBS Letters*, **208**, 413-417.

DROBAK, B.K., ALLAN, E.F., COMERFORD, J.G., ROBERTS, K., and DAWSON, A.P., 1988. Presence of guanine-nucleotide binding proteins in a plant hypocotyl microsomal fraction. *Biochemical and Biophysical Research Communications*, **150**, 899-903.

EINSPAHR, K.J., and THOMPSON, G.A., 1990. Transmembrane signalling via phosphatidylinositol 4,5-bisphosphate hydrolysis in plants. *Plant Physiology*, **93**, 361-366.

FINAN, P.M., WHITE, I.R., FINDLAY, J.B.C., and MILLNER, P.A., 1991. Identification of nucleoside diphosphate kinase from pea microsomal membranes. *Biochemical Society Transactions*, **20**, in press.

GOLDSMITH, P., GIERSCHIK, P., MILLIGAN, G., UNSON, C.G., VINITSKY, R., MALECH, H.L., and SPIEGEL, A.M., 1987. Antibodies directed against synthetic peptides distinguish between GTP-binding proteins in neutrophil and brain. *Journal of Biological Chemistry*, **262**, 14683-14688.

HASUNUMA, K., and FUNADERA, K., 1988. GTP-binding protein(s) in green plant, *Lemna paucicosta*. *Biochemical and Biophysical Research Communications*, **143**, 908-912.

HASUNUMA, K., FURUKAWA, K., TOMITA, K., MUKAI, C., and NAKAMURA, T., 1988. GTP-binding proteins in etiolated epicotyls of *Pisum sativum* (Alaska) seedlings. *Biochemical and Biophysical Research Communications*, **148**, 133-139.

JACOBS, M., THELEN, M.P., FARNDALE, R.W., ASTLE, M.C., and RUBERY, P., 1988. Specific guanine nucleotide binding by membranes from *Cucurbita pepo* seedlings. *Biochemical and Biophysical Research Communications*, **155**, 1478-1484.

KAZIRO, Y., ITOH, H., KOZASA, T., NAKAFUKU, M., and SATOH, T., 1991. Structure and function of signal transducing GTP-binding proteins. *Annual Review of Biochemistry*, **60**, 349-400.

KIKKAWA, S., TAKAHASHI, K., TAKAHASHI, K.-I., SIMADA, N., UI, M., KIMURA, N., and KATADA, T., 1990. Conversion of GDP into GTP by nucleoside diphosphate kinase on the GTP binding proteins. *Journal of Biological Chemistry*, **265**, 21536-21540.

KJELLBOM, P., and LARSSON, C., 1984. Preparation and polypeptide composition of chlorophyll free plasma membranes from leaves of light grown spinach and barley. *Physiologia Plantarum*, **62**, 501-509.

LACHMANN, P.J., STRANGEWAYS, L., VYAKARNAM, A., and EVAN, G., 1986. Raising antibodies by coupling peptides to PPD and immunising BCG-sensitised animals. *CIBA Symposium*, **119**, 25-37.

MA, H., YANOFSKY, M.F., and MEYEROWITZ, E.M., 1990. Molecular cloning and characterisation of *GPA1*, a G-protein α subunit gene from *Arabidopsis thaliana*. *Proceedings of the National Academy of Sciences, USA*, **87**, 3821-3825.

MILLNER, P.A., 1987. Are guanine nucleotide-binding proteins involved in regulation of the thylakoid protein kinase? *FEBS Letters*, **226**, 155-160.

MUMBY, S.M., KAHN, R.A., MANNING, D.R., and GILMAN, A.G., 1986. Antisera of designed specificity for subunits of guanine nucleotide-binding regulatory proteins. *Proceedings of the National Academy of Sciences, USA*, **83**, 265-269.

SCHLOSS, J.A., 1990. A *Chlamydomonas* gene encodes a G-protein β subunit like polypeptide. *Molecular and General Genetics*, **221**, 443-452.

WISE, A., and MILLNER, P.A., 1991. Evidence for the presence of GTP-binding proteins in tobacco leaf and maize hypocotyl plasmalemma. *Biochemical Society Transactions*, **20**, in press.

ZBELL, B., SCHWENDEMANN, I., and BOPP, M., 1989. High affinity GTP-binding on microsomal membranes prepared from moss protonema of *Funaria hygrometrica*. *Journal of Plant Physiology*, **134**, 639-641.

Protein Targeting and Assembly in Membranes

Introduction

The final two chapters of this book deal with the way in which proteins, having been synthesised, are directed to their correct locations within the cell. Of all the subjects covered in this volume, this topic is the least understood, particularly in plants. However, it presents a most challenging and exciting dimension to plant biochemistry and should provide new insights into cellular membrane biogenesis.

It has been established that the pathway from the lumen of the endoplasmic reticulum (ER) to the Golgi and thence to the plasma membrane (PM) is the 'default' pathway. Within the lumen of the ER, proteins which are destined for the other compartments interact with ER proteins such as BiP and PDI, the former being a chaperone protein of the HSP-70 family. The role of these chaperones in targeting is not clear, but a function in refolding or correct assembly of proteins is likely; with correct oligomerisation being required for exit from the ER.

Targeting to the ER requires a signal sequence at the N-terminus of a protein. Translocation across the ER is a co-translational process involving interaction between the nascent N-terminus signal sequence and the signal recognition particle. This, in turn, interacts with the surface of the rough ER.

A fundamental difference between protein translocation across the ER and translocation across other organellar membranes is that the latter is post-translational. Therefore, this requires that the cytoplasmically synthesised proteins destined for the mitochondria or chloroplasts are kept in a translocationally competent form (ie. unfolded or unassembled). This state is maintained by interactions with cytoplasmic chaperones such as the HSP-70 family of proteins.

Possibly because it is the least difficult to work with *in vitro*, protein targeting to the chloroplast is the best understood targeting system in plants. However, it is very different from targeting to the ER, in that cytoplasmically synthesised precursor proteins (polypeptides with N-terminal extensions called the transit sequence) interact with putative outer envelope receptors. Proteins are then translocated across the envelope and into the stroma where the chloroplast transfer domain of the transit sequence is removed. Proteins destined for the thylakoid membrane usually contain further targeting information, either in the form of a cleavable thylakoid transfer domain in the N-terminus of the transit sequence, eg. OEC-22,33, or a non-cleavable signal in the C-terminus of the matrix protein, eg. LHC-II.

In perhaps an analogous manner to proteins crossing the cytoplasm, the polypeptides in the stroma need to maintain a 'translocatable competence' and so interact with a number of stromal factors. However, it is not clear if all the thylakoid targeted proteins require such factors (Mould *et al*). It may be that the primary amino acid sequence of some proteins is more or less prone to malfolding.

This process of protein targeting to the chloroplast is described in more detail in the ensuing contributions.

Sequence Determinants for Protein Import into Chloroplasts and Thylakoid Membrane Protein Assembly

Gunnar von Heijne

Department of Molecular Biology
and Karolinska Institute Center for Structural Biochemistry
NOVUM
S-141 52, Huddinge
Sweden

Summary

Most nuclear encoded chloroplast proteins are targeted to the organelle by transient N-terminal extensions or transit peptides (cTPs). Transit peptides from higher plants are rich in hydroxylated residues but contain only few acidic residues; in contrast cTPS from *Chlamydomonas* contain many positively charged residues, and appear to be able to form amphiphilic α-helices. Upon import, cTPs are cleaved by a stromal processing peptidase. In many but not all cTPs the cleavage site is characterized by a weakly conserved motif: (I/V)-X-(A/C) $\downarrow$ A. Proteins destined for the lumen of the thylakoids have a second targeting signal placed C-terminally to the stromal transit peptide that has a structure very similar to the secretory signal peptides of prokaryotes. Finally, integral thylakoid membrane proteins have recently been shown to have a preponderance of positively charged residues in their stromafacing domains, and thus follow the same "positive inside-rule" that characterises both prokaryotic and eukaryotic plasma membrane proteins. It thus seems likely that the processes of translocation across, or integration into, the thylakoid membrane are mechanistically similar to the corresponding processes in bacteria.

1. Introduction

From the point of view of protein targeting, chloroplasts provide an exceptionally interesting system. With its two envelope membranes and internal thylakoid stacks, it presents at least six distinct sub-organellar compartments among which an incoming protein

Transport and Receptor Proteins of Plant Membranes
Edited by D.T. Cooke and D.T. Clarkson, Plenum Press, New York, 1992

195

has to navigate. In a general sense, plant cells have solved this problem in the same way that they have solved the problems of routing proteins into the mitochondria or into the secretory pathway: by sticking address-labels on the nascent proteins and furnishing the organelle with appropriate receptor- and translocator-proteins that recognise the address label and haul the protein into the organelle.

This review will focus on the address labels, or transit peptides (cTPs) as they are properly called, for chloroplast import and further sorting within the organelle. First, the discussion will focus on stroma-targeting cTPs, followed by thylakoid transfer domains, and finally the sequence characteristics of the thylakoid membrane proteins that determine their membrane topology.

2.   Stroma-Targeting Transit Peptides

The amino acid sequences of close to 300 cTPs are presently known, and, thanks to a joint effort of laboratories in the US, Germany, and Sweden, these sequences have been collected into a simple database called CHLPEP (von Heijne, Hirai, Klösgen, Steppuhn, Bruce, Keegstra and Herrmann, 1991). Such collections are invaluable when one attempts to find conserved characteristics among cTPs that might serve as the basic elements recognised by the import receptor(s). A quick glance at Figure 1 will convince the reader that one outstanding attribute of cTPs is their amazing sequence variability; nevertheless, these sequences share certain notable features. In terms of overall amino acid composition, there is a clear enrichment of hydroxylated residues and a corresponding lack of acidic residues in cTPs from higher plants (von Heijne, Steppuhn and Herrmann, 1989). Basic residues are found at frequencies comparable to those in soluble proteins in general, but not all cTPs contain basic amino acids. This is in contrast to mitochondrial targeting peptides (mTPs), where basic residues abound (von Heijne, 1986b) and seem to be of critical functional importance (von Heijne, 1990).

Closer scrutiny reveals that the N-terminal 5-10 residues rarely contain charged residues or prolines (von Heijne, et al., 1989) and often start Met-Ala. The central part presents no obvious residue-patterns beyond the overall compositional preferences noted above; in particular, there is no indication of amphiphilic segments like those found in mTPs. The region immediately surrounding the cleavage-site seems slightly more conserved, and there are some hints that this region may form an amphiphilic β-strand (von Heijne, et al., 1989). There are often one or two arginines some 5 - 10 residues upstream of the cleavage-site, and a semi-conserved (I/V)-X-(A/C) ↓ A has been found in about two-thirds of the known cTPs (Gavel and von Heijne, 1990).

Thus, there is very little to suggest any particular secondary or tertiary structure for higher plant cTPs. The high incidence of hydroxylated residues and prolines rather seems to indicate a largely unfolded conformation, and we have recently proposed the rather wild hypothesis that cTPs may in fact have been designed to *avoid* all strong conformational preferences (von Heijne and Nishikawa, 1991), possibly to be able to interact efficiently with cytosolic and chloroplast-specific chaperones that might form part of the targeting machinery.

However, cTPs from the green alga *Chlamydomonas* have a very different structure. In terms of both overall amino acid composition and amphiphilic properties they are much more similar to mTPs than to higher plant cTPs (Franzèn, Rochaix and von Heijne, 1990). Thus, they have a high content of basic residues (arginines in particular), and have a high potential for forming amphiphilic α-helices. This difference between green algae and higher plant cTPs may explain an early observation that cTPs could mis-target a passenger protein into yeast mitochondria (Hurt, Soltanifar, Goldschmidt-Clermont and Schatz, 1986), since the cTP in question was from a *Chlamydomonas* protein. More recent data suggest that higher plant cTPs and mTPs are able to discriminate quite efficiently between mitochondria and chloroplasts (Boutry, Nagy, Poulsen, Aoyagi and Chua, 1987; Whelen, Knorpp and Glaser, 1990).

3.   Thylakoid Transfer Domains

Already when the first sequences of nuclear encoded thylakoid lumen proteins became available, it was realised that their targeting peptides had a bipartite structure with a typical stroma-targeting cTP immediately followed by a stretch of apolar amino acids and a cleavage site region highly reminiscent of a prokaryotic secretory signal peptide (Smeekens, van Binsbergen and Weisbeek, 1985; von Heijne, *et al.*, 1989), Figure 1. This bipartite structure seems to be reflected in a two-step targeting process: first, the precursor is imported into the stromal compartment and the stroma-targeting domain is removed by the stromal processing peptidase (Smeekens, Bauerle, Hageman, Keegstra and Weisbeek, 1986), but see also (Bauerle, Dorl and Keegstra, 1991; Bauerle and Keegstra, 1991); then, the signal peptide-like thylakoid transfer domain (TTD) serves to target the protein further to the thylakoid import system, and the TTD is finally removed by a thylakoid processing peptidase (Hageman, Robinson, Smeekens and Weisbeek, 1986). The substrate specificity of the thylakoid processing peptidase is very similar but apparently more restricted than that of the prokaryotic leader peptidase enzyme (Halpin, Elderfield, James, Zimmermann, Dunbar and Robinson, 1989; von Heijne, *et al.*, 1989; Shackleton and Robinson, 1991), with a conserved Ala-X-Ala ↓ motif immediately preceding the cleavage site.

4.   Thylakoid Membrane Protein Assembly

The close mechanistic similarity between thylakoid protein import and protein secretion in bacteria is of course not very surprising in view of the endosymbiont origin of the chloroplast. Further, it suggests that the mechanism of protein assembly into the thylakoid membrane may also be similar to that for inner membrane bacterial proteins. In *E. coli*, membrane protein assembly has recently been shown to be guided by an interplay between two kinds of sequence elements: apolar segments of about 20 residues length that form transmembrane α-helices in the final structure, and positively charged arginine and lysine residues in the regions flanking the apolar stretches (von Heijne, 1988). The first hints came from statistical sequence analyses, which showed that arginines and lysines are up to four-fold more frequent in cytoplasmic as compared to periplasmic segments (von Heijne, 1986a); this "positive inside-rule" (von Heijne and Gavel, 1988) has since been given a strong experimental underpinning from studies where it was shown that the addition or removal of positively charged residues in strategic locations can "flip" the orientation of integral membrane proteins; see (Boyd and Beckwith, 1990; Dalbey, 1990) for up-to-date reviews.

We have recently found that thylakoid membrane proteins also follow the positive inside-rule (Gavel, Steppuhn, Herrmann and von Heijne, 1991), with more positively charged residues in stromal (non-translocated) as compared to lumenal (translocated)

A

Triticum aestivum
Fructose-1,6-bisphosphatase (stroma)

MAAATTTTSRPLLLSRQQAAASSLQCRLPRRPGSSLFAGQGQASTPNVRCMAVVDTASAPAPAAARKRSSY
DMITLTT

Acetabularia mediterranea
SSU (stroma)

MASIMMNKSVVLSKECAKPLATPKVTLNKRGFATTIATKNRE↓MMVWQPFNNKMFETFSFLPP

B

Chlamydomonas reinhardtii
SSU (stroma)

MAVIAKSSVAAVARPARSSVRPMAALKPAVKAAPVAAPAEAND↓MMVWTPVNNK

Chlamydomonas reinhardtii
OEE2 (thylakoid lumen)

MATALCNKAFAAAPVARPASRRSAVVVRASGSDVSRRAALR<u>GFAGAAALV</u>SSSPANA↓
AYGDSANVFGKVTNKSGFVP

C

Silene pratensis
Plastocyanin (thylakoid lumen)

MATVTSSAAVAIPSFAGLKASSTTRAATVKVAVATPRMSIKASLKD<u>VGVVVAATAAAGILAG</u>NAMA↓AEVL
LGSSDGGLAFVPSDLS

Spinacia oleracea
16 kd protein (thylakoid lumen)

MAQAMASMAGLRGASQAVLEGSLQISGSNRLSGFℓTSRVAVPKMGLNIRAQQVSAEAETSRR<u>AMLGFVAAG</u>
<u>LASGSFV</u>KAVLA↓EARPIVVGPPPPLSGGLPGT

Figure 1 A collection of stromal cTPs from higher plants (A) and *Chlamydomonas* (B), and thylakoid targeting (C). The apolar segments of the thylakoid transfer domains are underlined. Cleavage sites are marked by "↓". All sequences were extracted from the CHLPEP database (von Heijne, *et al.*, 1991).

segments. This suggests that positively charged amino acids may play a similar role as determinants of membrane protein topology in both bacteria and chloroplasts.

Acknowledgement

This work was supported by grants from the Swedish Natural Sciences Research Council and the Swedish National Board for Technical Development.

REFERENCES

BAUERLE, C., DORL, J., and KEEGSTRA, K., 1991. Kinetic analysis of the transport of thylakoid lumenal proteins in experiments using intact chloroplasts. *Journal of Biological Chemistry*, **266**, 5884-5890.
BAUERLE, C., and KEEGSTRA, K., 1991. Full-length plastocyanin precursor is translocated across isolated thylakoid membranes. *Journal of Biological Chemistry*, **266**, 5876-5883.
BOUTRY, M., NAGY, F., POULSEN, C., AOYAGI, K., and CHUA, N.H., 1987. Targeting of bacterial chloramphenicol acetyltransferase to mitochondria in transgenic plants. *Nature*, **328**, 340-342.
BOYD, D., and BECKWITH, J., 1990. The role of charged amino acids in the localisation of secreted and membrane proteins. *Cell*, **62**, 1031-1033.

DALBEY, R.E., 1990. Positively charged residues are important determinants of membrane protein topology. *Trends Biochemical Science*, **15**, 253-257.

FRANZÉN, L.G., ROCHAIX, J.D., and VON HEIJNE, G., 1990. Chloroplast transit peptides from the green alga *Chlamydomonas-reinhardtii* share features with both mitochondrial and higher plant chloroplast presequences. *FEBS Letters*, **260**, 165-168.

GAVEL, Y., STEPPUHN, J., HERRMANN, R., and VON HEIJNE, G., 1991. The "Positive-Inside" rule applies to thylakoid membrane proteins. *FEBS Letters*, **282**, 41-46.

GAVEL, Y., and VON HEIJNE, G., 1990. A conserved cleavage-site motif in chloroplast transit peptides. *FEBS Letters*, **261** 455-458.

HAGEMAN, J., ROBINSON, C., SMEEKENS, S., and WEISBEEK, P., 1986. A thylakoid processing protease is required for complete maturation of the lumen protein plastocyanin. *Nature*, **324**, 567-9.

HALPIN, C., ELDERFIELD, P.D., JAMES, H.E., ZIMMERMANN, R., DUNBAR, B., and ROBINSON, C., 1989. The reaction specificities of the thylakoidal processing peptidase and *Escherichia-coli* leader peptidase are identical. *EMBO Journal*, **8**, 3917-3921.

HURT, E.C., SOLTANIFAR, N., GOLDSCHMIDT-CLERMONT, M., and SCHATZ, G., 1986. The cleavable pre-sequence of an imported chloroplast protein directs attached polypeptides into yeast mitochondria. *EMBO Journal*, **5**, 1343-1350.

SHACKLETON, J.B., and ROBINSON, C., 1991. Transport of proteins into chloroplasts: The thylakoidal processing peptidase is a "signal" - type peptidase with stringent substrate requirements at the -3 and -1 positions. *Journal of Biological Chemistry*, **266**, 12152-12156.

SMEEKENS, S., BAUERLE, C., HAGEMAN, J., KEEGSTRA, K., and WEISBEEK, P., 1986. The role of the transit peptide in the routing of precursors toward different chloroplast compartments. *Cell*, **46**, 365-75.

SMEEKENS, S., VAN BINSBERGEN, J., and WEISBEEK, P., 1985. The plant ferredoxin precursor: nucleotide sequence of a full length cDNA clone. *Nucleic Acids Research*, **13**, 3179-94.

VON HEIJNE, G., 1986a. The distribution of positively charged residues in bacterial inner membrane proteins correlates with the transmembrane topology. *EMBO Journal*, **5**, 3021-27.

VON HEIJNE, G., 1986b. Mitochondrial targeting sequences may form amphiphilic helices. *EMBO Journal*, **5**, 1335-42.

VON HEIJNE, G., 1988. Transcending the impenetrable: how proteins come to terms with membranes. *Biochim Biophysics Acta*, **947**, 307-33.

VON HEIJNE, G., 1990. Protein Targeting Signals. *Current Opinions in Cell Biology*, **2**, 604-608.

VON HEIJNE, G., and GAVEL, Y., 1988. Topogenic signals in integral membrane proteins. *European Journal of Biochemistry*, **174**, 671-8.

VON HEIJNE, G., HIRAI, T., KLÖSGEN, R.-B., STEPPUHN, J., BRUCE, B., KEEGSTRA, K., and HERRMANN, R., 1991. CHLPEP - A database of chloroplast transit peptides. *Plant Molecular Biology Reporter*, **9**, 104-126.

VON HEIJNE, G., and NISHIKAWA, K., 1991. Chloroplast transit peptides - The perfect random coil? *FEBS Letters*, **278**, 1-3.

VON HEIJNE, G., STEPPUHN, J., and HERRMANN, R.G., 1989. Domain structure of mitochondrial and chloroplast targeting peptides. *European Journal of Biochemistry*, **180**, 535-545.

WHELAN, J., KNORPP, C., and GLASER, E., 1990. Sorting of precursor proteins between isolated spinach leaf mitochondria and chloroplasts. *Plant Molecular Biology*, **14**, 977-982.

Mechanism and Energetics of Protein Transport Across the Thylakoid Membrane

Ruth M. Mould, Julie W. Meadows, Jamie B. Shackleton
and Colin Robinson

Department of Biological Sciences
University of Warwick
Coventry, CV4 7AL, UK

1.   Introduction

The assembly of the photosynthetic machinery in the higher plant chloroplasts is a complex process, in part because the component proteins are synthesised by two distinct genetic systems. Approximately half of the known thylakoidal proteins are synthesised within the organelle, whereas the remainder are imported after synthesis in the cytosol. The biogenesis of the latter group of proteins has attracted considerable interest in recent years, since these proteins must cross both of the envelope membranes and the soluble stromal phase to reach the thylakoid membrane. Of particular complexity is the import of hydrophilic thylakoid lumen proteins, which must in addition traverse the thylakoid membrane to reach their site of function.

Studies on the import of thylakoid lumen proteins have focused primarily on four prominent photosynthetic proteins: plastocyanin, a mobile electron carrier, and the 33kDa, 23kDa and 16kDa proteins (33K,23K,16K) of the photosystem II oxygen-evolving complex. It is now believed that these proteins are imported by a two-step mechanism after synthesis in the cytosol. The proteins are initially synthesised with bipartite pre-sequences which consist of two targeting signals in tandem, specifying "envelope transit" and "thylakoid transfer". The transit sequence first directs transport into the stroma after which it is removed by a stromal processing peptidase (SPP). The transfer signal then mediates transport of the intermediate form across the thylakoid membrane, after which processing to the mature size is carried out by a thylakoidal processing peptidase, TPP (Hageman, Robinson, Smeekens and Weisbeek, 1986; James, Bartling, Musgrove, Kirwin, Herrmann, and Robinson, 1989; Ko and Cashmore, 1989). This two-stage import model, which is depicted in Figure 1, was prompted by three key observations.

(1)   The pre-sequences of lumenal proteins contain two structurally distinct domains. The aminoterminal domains clearly resemble the pre-sequences of stromal proteins,

being hydrophilic, basic and rich in hydroxylated residues. In contrast, the carboxy terminal thylakoid transfer domains contain a hydrophobic core region reminiscent of so-called signal sequences (von Heijne, Steppuhn, and Herrmann, 1989; Halpin, Elderfield, Musgrove, Dunbar, Zimmermann, and Robinson, 1989).

(2) Transient, intermediate-size stromal forms of several lumenal proteins have been observed during the import of *in vitro* synthesised precursors into isolated chloroplasts (Smeekens, Bauerle, Hagemann, Keegstra, and Weisbeek, 1986; James *et al.*, 1989).

(3) The two-step proteolytic maturation sequence can be reconstructed *in vitro* by the addition of the appropriate processing peptidases. Both SPP and TPP have been highly purified (Robinson and Ellis, 1984; Kirwin, Elderfield and Robinson, 1987) and shown to process lumenal protein precursors to intermediate and mature sizes, respectively (Hageman *et al.*, 1986; James *et al.*, 1989).

The above observations strongly suggested that the import of thylakoid lumen proteins involved the operation of two distinct protein translocation systems, one of which is located in the chloroplast envelope and the other in the thylakoid membrane. In order to understand in greater detail the mechanism of the thylakoid system, we have sought to

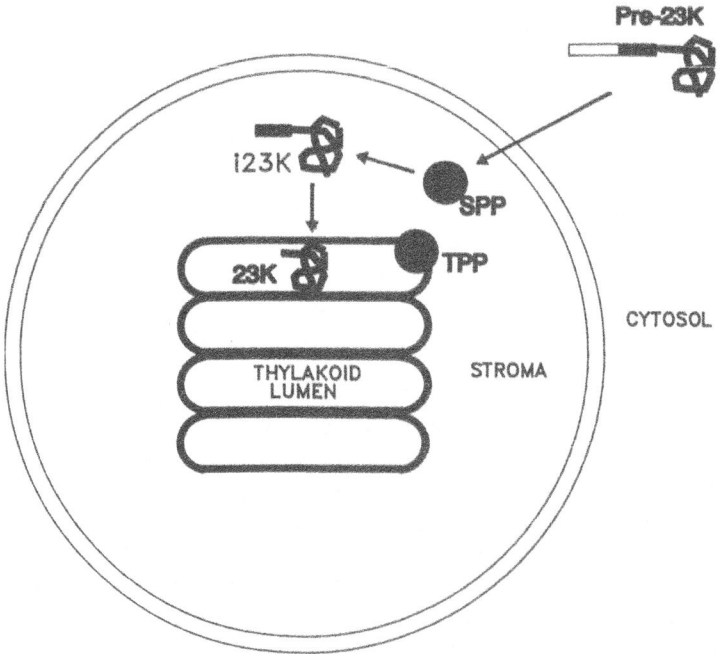

Figure 1 Two-step model for the import of thylakoid lumen proteins.
Lumenal proteins such as the 23kDa protein of the oxygen-evolving complex (23K) are synthesised in the cytosol with a bipartite pre-sequence. After synthesis, pre-23K is imported into the stroma and cleaved to an intermediate form (int-23K) by a stromal processing peptidase (SPP). i-23K is subsequently transported across the thylakoid membrane and processed to the mature size by a thylakoidal processing peptidase, TPP.

develop an efficient assay for the import of proteins into isolated thylakoids. Initially, it was found that 33K could be imported into pea thylakoids with reasonable efficiency in the presence of ATP. However, a second lumenal protein, 23K, was imported with very low efficiency in this assay system. More recently, we have developed a light-driven assay for the efficient import of several lumenal proteins, and in this article we describe some of the features of the thylakoidal transport system which have emerged from these studies.

2. Light-Driven Protein Import into Isolated Thylakoids

Figure 2 shows the effects of light and added stromal extract on the import of 33K by isolated thylakoids. Pre-33K was synthesised by *in vitro* transcription-translation of a full-length cDNA and incubated with washed pea thylakoids. Lanes 1 to 4 show that, in the light, the inclusion of increasing levels of stromal extract leads to an increase in the levels of an intermediate-size 33K (i33K); this is due to the presence of SPP in the stromal

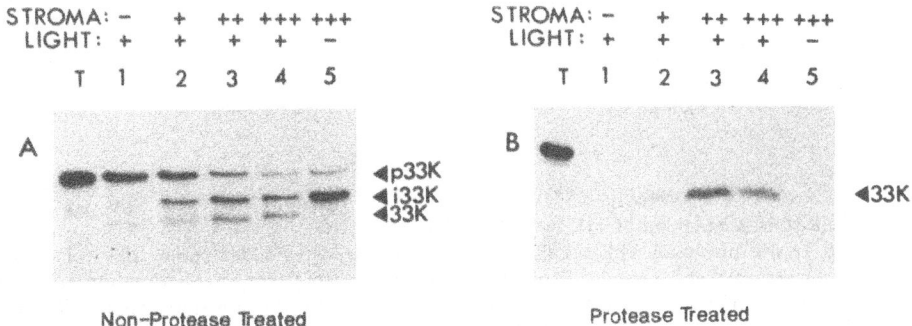

Non–Protease Treated   Protease Treated

Figure 2 Light-stimulated import of 33K by isolated thylakoids.
Pre-33K (lanes T) was incubated with pea thylakoids in the absence of added stromal extract (lanes 1) or in the presence of crude stromal extract in 10 mM HEPES-KOH, pH 8.0 at protein concentrations of 0.05, 0.2 and 1 mg ml^{-1} (lanes 2-4). Import assays were carried out under illumination using thylakoids equivalent to 0.75 mg ml^{-1} chlorophyll per incubation mixture. Lane 5, incubation conditions were as in lane 4, except that incubation was varied out in the dark. Samples were analysed directly after incubation (upper panel) or after protease treatment (lower panel) of the thylakoids.

extract. In addition, there is an increase in the levels of mature-size 33K produced. The mature-size 33K is resistant to protease-treatment of the thylakoids, demonstrating that import into the lumenal space has taken place. Lane 5 shows that in the presence of stromal extract, but in the dark, pre-33K is efficiently processed to i33K but essentially no import takes place. These data indicate that the import of 33K into thylakoids requires both light and stromal extract.

Import assays using 23K as a substrate are shown in Figure 3. In lanes 1-3, pre-23K was incubated with thylakoids in the light and in the presence of increasing levels of stroma. As expected, higher levels of stromal extract produce a greater conversion of pre-

23K, the intermediate form. However, protease-resistant mature-size 23K is generated efficiently in both the presence and absence of stromal extract. In the dark (lane 4) no import is observed.

The data shown in Figures 2 and 3 demonstrate that light is essential for the efficient import of 23K and 33K into thylakoids, but that the two proteins differ notably in their requirements for stromal extract. 23K is imported in the complete absence of stroma demonstrating that the generation of the intermediate form by SPP is not a strict requirement for transport into the thylakoid lumen. Clearly, the full precursor protein can be efficiently recognised and imported by the thylakoid protein translocation system. In contrast, import of pre-33K requires concentrated stromal extract and we considered it likely that this finding indicated that the generation of i33K by SPP in the extract was a prerequisite for subsequence transport across the thylakoid membrane. However, these data also raised the possibility that a stromal factor other than SPP is required for the import of 33K. In order

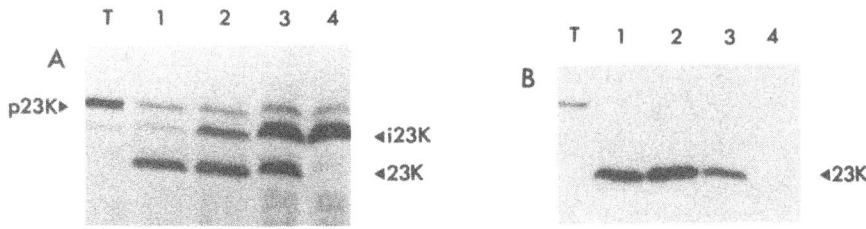

Figure 3 Light-stimulated import of 23K by isolated thylakoids.
Pre-23K (lane T) was incubated with pea thylakoids in the absence of added stroma (lanes 1) or in the presence of stromal extract at 0.2 mg ml⁻¹ (lanes 2) or 1 mg ml⁻¹ protein (lanes 3). Mixtures were illuminated for 20 min. Lanes 4, incubation conditions were as in lanes 3 except that the mixture was kept in the dark. Samples were analysed directly after incubation (upper panel) or after protease treatment of the thylakoids p23K, pre-23K.

to resolve these possibilities, an artificial i33K was synthesised in which the first, envelope transit, sequence and SPP cleavage site were missing. Bassham, Bartlin, Mould, Dunbar, Weisbeek, Herrmann, and Robinson (in press) identified the sites at which SPP cleaves within several lumenal protein precursors, including that of 33K. Using this information, the last residue of the 33K envelope transit sequence was changed to methionine by site-specific mutagenesis of the CDNA and the initiation codon was deleted. This produced a cDNA sequence which, when transcribed and translated gave rise to an intermediate size 33K (i33K) which was one residue larger than the authentic intermediate and which was no longer recognised by SPP. Figure 4 shows thylakoid import assays using this artificial i33K; import is observed in the presence but not in the absence of stromal extract. We conclude from these data that at least one stromal factor, other than SPP, is required for the import of 33K, but not 23K, into isolated thylakoids. Other work has shown that the factor is heat-sensitive and non-dialysable, which strongly suggests that the factor is proteinaceous. Future studies will aim to determine the structure and mode of action of the factor.

3. Energetics of Protein Translocation Across the Thylakoid Membrane

Until recently, for largely technical reasons, very little was known about the mechanism of protein transport across the thylakoid membrane. Using the standard intact chloroplast import assay, it is very difficult to analyse in detail protein translocation across the internal membrane network. However, the data in Figures 2 and 3, using the thylakoid import assay, showed that the presence of light is a critical requirement for efficient import of both 23K and 33K by isolated thylakoids. It seemed likely that this reflected a role, either directly or indirectly, of thylakoidal electron transport in promoting protein translocation

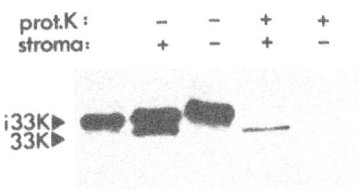

Figure 4 Import of an artificial i33K into pea thylakoids.
i33K was incubated with isolated pea thylakoids in the presence (+) or absence (-) of stromal extract. After incubation samples were analysed directly or after proteinase K treatment of the thylakoids.

and experiments were carried out to test this possibility. Import assays were done in the presence of electron transport inhibitors (a combination of dichloro phenyldimethylurea (DCMU) and methyl viologen) to block formation of the thylakoidal proton motive force (pmf). Other assays were carried out in the presence of valinomycin (which selectively dissipates the electrical potential component, $\Delta\psi$ of the pmf) and nigericin, which collapses the proton gradient component, ΔpH. Figure 5 shows the result of a thylakoid import assay using pre-23K as a substrate. In the presence of stroma and light, pre-23K is converted to the intermediate form (by SPP in the stromal extract) and also to the mature size (lane 1). When this mixture is protease-treated after incubation (lane 5) the mature-size 23K is protected from digestion, showing that these molecules have been imported into the lumen. In the presence of DCMU/methyl viologen, pre-23K is efficiently processed to the intermediate form but no mature-size polypeptide appears, showing that import has been completely inhibited (lanes 2 and 6). The presence of valinomycin (lanes 3 and 7) leads to a very slight inhibition of import, but nigericin again blocks import completely (lanes 4 and 8). We conclude from this experiment that protein transport across the thylakoid membranes requires an energised membrane and that the dominant component driving the translocation process is the proton gradient and not the electrical potential of the total protonmotive force. It remains to be determined precisely how the pmf drives protein transport across the thylakoid membrane.

4. Discussion

The primary aims of this study were two-fold: (i) to develop an efficient *in vitro* assay for thylakoidal protein import and, (ii) to understand the mechanism of the overall translocation process. The first of these objectives appears to have been reached, in that it is clearly possible to isolate pea thylakoids which are highly competent for the import of 33K and 23K. The efficiencies of import of 23K and 33K are ca. 60% and 40%, respectively, of available precursor, which is comparable with other *in vitro* import systems. Recent evidence (obtained in collaboration with R.G. Herrmann, Munich) has indicated that the third component of the oxygen-evolving complex, 16K, can also be imported by isolated thylakoids. Import of this protein is even more efficient (at least 90% of available precursor) and is also driven by light.

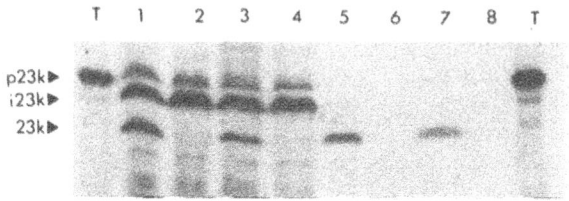

Figure 5 Light-dependent import of 23K by isolated thylakoids.
Pre-23K was incubated with stroma and thylakoids in the light (lane 1) and in the presence of DCMU methyl viologen (lane 2), valinomycin (lane 3) or nigericin (lane 4). Lanes 5-8; as in lanes 1-4 except that the thylakoids were protease-treated after incubation. Lane T, translation product; p23K, i23K, precursor and intermediate forms of 23K.

In the course of developing this import assay, several features of the translocation mechanism have emerged. The first of these concerns the energetics of this process; it is now clear that the transport of 33K, 23K and 16K across the thylakoid membrane is driven by a proton gradient. A major priority in future work will be to determine how the thylakoidal proton motive force is harnessed for this purpose, since currently this process is obscure. In addition, further studies are required in order to obtain a more complete understanding of the energetics of thylakoidal protein transport. For example, it is possible that this process requires ATP, as does protein transport across the chloroplast envelope membranes (Theg, Bauerle, Olsen, Selman, and Keegstra, 1989).

Further work is also required to resolve an interesting question raised by these findings: why do the transport energetics of 33K, 23K and 16K differ from those of plastocyanin? Although plastocyanin is imported into the thylakoid lumen by an apparently similar two-stage pathway the transport of this protein across the thylakoid membrane does not require a proton motive force (Theg *et al.*, 1989). The basis for this fundamental mechanistic difference is presently unclear, but these observations raise the possibility that more than one type of translocation system is operational in the thylakoid membrane.

In addition to the involvement of a proton gradient, our results have also shown that the import of 33K requires at least one stromal factor, presumably a protein. Again, though, it is intriguing that another lumenal protein (23K) does not require the presence of this factor. The basis for this difference is also unclear, but one possibility under consideration is that the factor is required to maintain 33K in an unfolded, import-competent confirmation. If so, it is possible that pre-23K is synthesised, at least *in vitro*, in a more loosely folded state which enables it to bypass any requirement for this factor.

REFERENCES

BASSHAM D.C., BARTLING, D., MOULD, R.M., DUNBAR, B., WACHTER, E., WEISBEEK, P., HERRMANN, R.G., and ROBINSON, C., 1992. Transport of proteins into chloroplasts. Delineation of envelope "transit" and thylakoid "transfer" signals within the pre-sequences of three imported thylakoid lumen proteins. *Journal of Biological Chemistry*, (in press).

HAGEMAN, J., ROBINSON, C., SMEEKENS, S., and WEISBEEK, P., 1986. A thylakoid processing peptidase is required for complete maturation of the lumen protein plastocyanin. *Nature*, **324**, 567-569.

HALPIN, C., ELDERFIELD, P.D., JAMES, H.E., ZIMMERMANN, R., DUNBAR, B., and ROBINSON, C., 1989. The reaction specificities of the thylakoidal processing peptidase and *Escherichia coli* leader peptidase are identical. *EMBO Journal*, **8**, 3917-3922.

JAMES, H.E., BARTLING, D., MUSGROVE, J.E., KIRWIN, P.M., HERRMANN, R.G., and ROBINSON, C., 1989. Transport of proteins into chloroplasts. Import and maturation of precursors to the 33, 23 and 16kDa proteins of the oxygen-evolving complex. *Journal of Biological Chemistry*, **264**, 19573-19576.

KIRWIN, P.M., ELDERFIELD, P.D., and ROBINSON, C., 1987. Transport of proteins into chloroplasts. Partial purification of a thylakoidal processing peptidase involved in plastocyanin biogenesis. *Journal of Biological Chemistry*, **262**, 16386-16390.

KO, K., and CASHMORE, A.R., 1989. Targeting of proteins to the thylakoid lumen by the bipartite transit peptide of the 33kd oxygen-evolving protein. *EMBO Journal*, **8**, 3187-3194.

ROBINSON, C., and ELLIS, R.J., 1984. Transport of proteins into chloroplasts. Partial purification of a chloroplast protease involved in the processing of imported precursor polypeptides. *European Journal of Biochemistry*, **142**, 337-342.

SMEEKENS, S., BAUERLE, C., HAGEMAN, J., KEEGSTRA, K., and WEISBEEK, P., 1986. The role of the transit peptide in the routing of precursors toward different chloroplast compartments. *Cell*, **46**, 365-375.

THEG, S., BAUERLE, C., OLSEN, L., SELMAN, B., and KEEGSTRA, K., 1989. Internal ATP is the only energy requirement for the translocation of precursor proteins across chloroplastic membranes. *Journal of Biological Chemistry*, **264**, 6730-6736.

VON HEIJNE, G., STEPPUHN, J., and HERRMANN, R.G., 1989. Domain structure of mitochondrial and chloroplast targeting peptides. *European Journal of Biochemistry*, **180**, 535-541.

INDEX

The manufacturer's authorised representative in the EU is Springer
Nature Customer Service Centre GmbH, Europaplatz 3, 69115 Heidelberg,
Germany. If you have any concerns regarding our products, please
contact ProductSafety@springernature.com

Printed and bound by CPI Group (UK) Ltd, Croydon, CR0 4YY

23/04/2026

02095629-0013